THE BIG BACKYARD

BACKYARD

THE SOLAR SYSTEM BEYOND PLUTO

RON MILLER

TWENTY-FIRST CENTURY BOOKS / MINNEAPOLIS

This book is dedicated to Annie Riker Ward-Capozzi

Twenty-First Century Books™
An imprint of Lerner Publishing Group, Inc.
241 First Avenue North
Minneapolis, MN 55401 USA

For reading levels and more information, look up this title at www.lernerbooks.com.

Main body text set in Adobe Garamond Pro.
Typeface provided by Adobe Systems.

Library of Congress Cataloging-in-Publication Data

Names: Miller, Ron, 1947– author.
Title: The big backyard : the solar system beyond Pluto / Ron Miller.
Description: Minneapolis, MN : Twenty-First Century Books, [2023] | Includes
    bibliographical references and index. | Audience: Ages 13–18 | Audience: Grades
    7–9 | Summary: "Deep space holds materials left over from the formation of
    the solar system. Astronomers have been making exciting discoveries on the
    outermost fringes and the New Horizons spacecraft brings new insights into the
    origins of the sun and planets"— Provided by publisher.
Identifiers: LCCN 2022024233 (print) | LCCN 2022024234 (ebook) |
    ISBN 9781728475349 (library binding) | ISBN 9781728485980 (ebook)
Subjects: LCSH: Trans-Neptunian objects—Juvenile literature. | Comets—Juvenile
    literature. | Kuiper Belt—Juvenile literature. | Pluto (Dwarf planet)—Juvenile
    literature.
Classification: LCC QB694 .M55 2023 (print) | LCC QB694 (ebook) | DDC
    523.49—dc23/eng20220826

LC record available at https://lccn.loc.gov/2022024233
LC ebook record available at https://lccn.loc.gov/2022024234

Manufactured in the United States of America
1-52087-50570-9/26/2022

CONTENTS

INTRODUCTION
VISITOR FROM AFAR

I n 2014 astronomers discovered a strange object beyond the orbit of Neptune, at nearly 29 astronomical units (AU). Since 1 AU is the average distance of Earth from the sun, or about 93 million miles (150 million km), that's about 2.7 billion miles (4.3 billion km) away. The strange object was large, probably between 62 and 93 miles (100 and 150 km) in diameter. Astronomers believe it's a large comet: a gigantic iceberg in space.

Named 2014 UN271, or Comet Bernardinelli-Bernstein for its discoverers, the object will make its closest approach to the sun in 2031. It will reach the orbit of Saturn—about ten times farther from the sun than Earth. If it is a comet, it will begin to develop a comet's distinctive coma and tail as the heat of the sun begins to evaporate the ice within it. It won't be visible to the naked eye, but astronomers will be able to observe it with telescopes.

People observe comets all the time, and many come much closer to Earth than 2014 UN271 will. Some are so close that they are easily visible in the sky. What makes this particular comet so interesting is where it came from.

Astronomers can calculate the orbit of 2014 UN271. The orbit is shaped like a very elongated ellipse, a stretched-out oval. At its farthest point from the sun, the orbit reaches 55,000 AU. That is nearly one light-year, the distance that light travels in one year: about 6 trillion miles (9.5 trillion km). It takes 2014 UN271 about 3 million years to make the long fall toward the sun and nearly 4.5 million years to return to where it started. This point is so far away that the nearest stars have almost as much of a gravitational pull on the comet as the sun has. The comet comes from the most distant frontier of our solar system, where the sun would be little more than a speck, more than thirty billion times dimmer than what we see in our sky on Earth.

What else might be at the farthest limits of the sun's reach? The comet 2014 UN271 is far from a unique visitor—a multitude of objects, some perhaps even more mysterious, lie in wait in the solar system's big backyard.

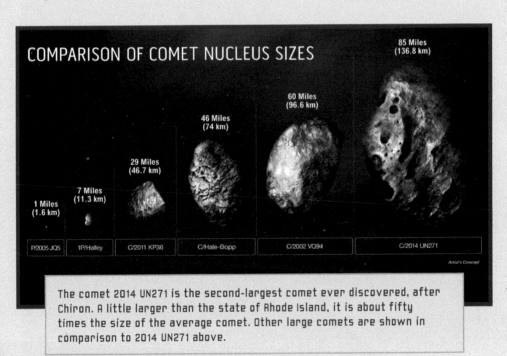

COMPARISON OF COMET NUCLEUS SIZES

85 Miles
(136.8 km)

60 Miles
(96.6 km)

46 Miles
(74 km)

29 Miles
(46.7 km)

7 Miles
(11.3 km)

1 Miles
(1.6 km)

| P/2005 JQ5 | 1P/Halley | C/2011 KP36 | C/Hale-Bopp | C/2002 VQ94 | C/2014 UN271 |

Artist's Concept

The comet 2014 UN271 is the second-largest comet ever discovered, after Chiron. A little larger than the state of Rhode Island, it is about fifty times the size of the average comet. Other large comets are shown in comparison to 2014 UN271 above.

CHAPTER 1

The Birth
of the Solar
System

To understand what lies at the outermost edges of our solar system, we need to go back to its very beginning, five billion years ago. Although no one can travel back in time, astronomers can use indirect evidence to understand how our solar system formed. They can observe solar systems forming around other stars, measure the materials found in our solar system, and run computer simulations based on the laws of physics. The dominant theory among astronomers is that the solar system formed from the gravitational collapse of a big cloud, or nebula. Instead of a sun or a system of planets, the cloud was made of organic molecules (which astronomers call dust), ice, and hydrogen gas. It was billions of miles wide and a few thousand times as massive as the sun.

The cloud was large enough that its gravity—as weak as it was—caused it to begin slowly collapsing. The molecules of gas and dust started drifting toward the center of the cloud, where there was more material and a slightly greater pull of gravity. As more material gathered at the center, the pull of gravity became stronger and the collapse became faster. Once the collapse started, it could not stop.

What started the collapse? What gave the cloud its initial nudge? It could have been almost anything—perhaps the shock wave from a nearby star exploding into a supernova. This would have acted in much the same way that a loud noise can start an avalanche. Scientists have found some evidence to support this theory. Measurements of the elements in the solar system reveal high concentrations of rare, radioactive elements. Supernova explosions are the only way to create these radioactive elements, so it is unlikely that the elements were already in the collapsing cloud. A nearby supernova is the most likely source. The presence of these radioactive elements is strong evidence that a supernova started the collapse of the nebula that eventually produced our solar system.

As the moving molecules drifted closer together, they occasionally bumped against one another. Each bump created a little heat, and the cloud grew warmer as it condensed.

You can observe this heating effect yourself with a bicycle pump. When you push down on the pump handle, the air inside the pump is compressed—it becomes denser. As this happens, the pump will grow warmer, which you can feel if you touch it. A tire will also grow warm as you inflate it, for the same reason: the air inside is being compressed and made denser.

The center of the cloud, where the gas and dust are densest, grew warm, then hot, as more and more material fell into it. As the center of the cloud became denser, it pulled even more gas and dust into it. The cycle increased rapidly. In just one year, a cloud of gas 2 trillion miles (3.2 trillion km) wide collapsed to only 200 million miles (322 million km)—ten thousand times smaller. Soon it became hot enough—over 3,000°F (1,650°C)—to vaporize any remaining particles of ice or dust. Meanwhile, the core of the cloud grew so dense and hot that it began to glow red, like a hot coal.

Hydrogen ionization occurs at a temperature of nearly 17,540°F (9,727°C). Driven by the increasing heat and pressure, the hydrogen atoms in the cloud's core collided with tremendous force. They collided so violently that their electrons were knocked out of their orbits. The hydrogen gas, no longer composed of intact atoms, became a cloud of charged particles called ions: the negatively charged electrons that had been knocked free, and the positively charged nuclei.

As the cloud's core continued to heat up, the nuclei begin to collide. Normally, two positively charged particles would repel each other, like the north poles of a pair of magnets. But as the core reached a temperature of 10,000,000°F to 20,000,000°F (5,500,000°C to 11,100,000°C), the extreme heat forced the nuclei not only to collide but to stick together. This nuclear fusion process created atoms of a new element—helium—one that was twice as heavy as the original hydrogen atom. Fusion doesn't just create heavier nuclei from lighter ones. Even more important, it releases energy in very large amounts.

The original cloud took about fifty million years to contract enough for its core to reach the temperature needed for fusion to begin. Until then, the collapsing cloud of gas was a protostar—it had the potential to become a star but wasn't quite one yet. But as soon as the spark of fusion was lit, it became a full-fledged star. Instead of glowing from the heat created by gravitational collapse, it shone from the light created by nuclear fusion. Fusion provided the star with a new source of energy, one much more powerful than one created by simple gravitational collapse. The outward pressure caused by the intense radiation balanced the inward pressure created by gravity. The process of gravitational collapse stopped.

And so, 4.5 billion years ago—50 million years after the original cloud of gas began to collapse—the star we call the sun was born. The force of the sun's radiation not only stopped the collapse of the star, but it also blew away most of the remaining dust and gas.

The young sun was a little smaller, fainter, and cooler than it is today. Now it is 10,000°F (5,538°C), about 200°F (111°C) hotter than it was at its birth. Because of its size, temperature, and other properties, astronomers consider our sun to be an average yellow dwarf star. They estimate that it has gone through about half of its hydrogen fuel and has about 6.5 billion years before it stops burning.

The early solar system formed in the middle of the solar nebula about 4.5 billion years ago.

Most stars you can see in the sky are main sequence stars, meaning that they are fusing hydrogen into helium in their cores. Main sequence stars can range in color and temperature. Red stars are cooler than blue stars. The sun is a little closer to the red end of the spectrum. After stars deplete their hydrogen reserves, they begin fusing helium into heavier elements and are no longer considered on the main sequence.

THE PLANETS

Not all the gas and dust in the original cloud wound up in the sun. Nor was it all blown away by radiation pressure. A lot was left over. All the excess material gathered in a protoplanetary disk, a rotating mass that looks a lot like a dark bun or pancake and often has a dimly glowing center. In 1992 the Hubble Space Telescope took photographs of newly formed stars in the Orion Nebula and found protoplanetary disks encircling them, providing the first direct evidence that protoplanetary disks are common and that our solar system likely formed in one.

Within the protoplanetary disk, particles of dust collided and stuck together, forming tiny clumps of material. As these clumps, or planetesimals, grew, they attracted more particles. This accretion process was similar to the way molecules of gas came together to form the sun. Most of these early collisions were relatively gentle, so the planetesimals tended to stick together instead of breaking into pieces. Small grains of dust slowly grew to the size of rocks, then boulders, and then asteroids miles across. For a single planetesimal in our solar system to grow from the size of a pinhead to a mountain may have taken only about one hundred thousand years. That's very quick by astronomical standards. The planetesimals continued to grow but at a slower rate—the dust and gas was being used up by the accreting objects, and the cloud was thinning.

As the planetesimals grew larger and the force of their gravity increased, they began to move faster and the collisions between them become more serious. Some of them shattered into pieces. A few of these planetesimals were large enough to survive these collisions, and they grew even larger, devouring the debris from the unluckier smaller objects. Once a body started accreting, it grew quickly. Earth may have grown from a grain of dust to nearly its present size in as few as forty million years.

But Earth was not the only lump that formed in the cloud. Many

As planetesimals crashed into one another during the planet formation process, they burned red hot.

others did. Like Earth, they continued to gather material as they circled the sun. The available material was not equally distributed throughout the disk. The sun's gravity pulled in the densest elements, such as iron and silicon, as well as many minerals. Lighter materials, such as hydrogen and water, remained in the outer parts of the disk. As a result, the parts of the disk closest to the sun had the densest supply of dust, ice, and other material. It was a little like throwing a mix of objects into a swimming pool: the denser items will sink to the bottom, while the lighter ones will float to the surface.

It was too warm near the sun for water to remain frozen, so the objects that formed near it were composed mostly of heavy metals and minerals. These objects eventually became the terrestrial planets: Mercury, Venus, Earth, and Mars. Farther out it was cold enough for permanent ice (ice that is never liquid) to form. Since most of the heavy materials had fallen in toward the sun, there was ten times as much ice as in the zone where the terrestrial planets formed. The gas giants—Jupiter, Saturn, Uranus, and Neptune—and their icy moons were born from the vast blizzard of snowflakes that orbited there.

Atacama Large Millimeter/submillimeter Array (ALMA), a radio telescope based in northern Chile, has produced many images of protoplanetary disks in various stages of development, confirming that stars and planet systems often form out of such disks. Glowing rings of gas and dust surround a central protostar. The gaps between the rings are regions where planetesimals have accreted material.

Beyond the gas giants, in the outermost region of the protoplanetary disk, it was cold enough for ice, but there was too little of it to form any very large objects. There we find the home of thousands of bodies that are mostly little more than giant icebergs: worlds like Pluto and its cousins in the Kuiper belt (a region of icy bodies that lies beyond the orbit of Neptune).

It took the solar system only about fifty million years after it started to form to sort itself out into its present familiar order of small, rocky planets near the sun and lighter, gas giant planets farther away.

A HISTORY OF DISCOVERY

The word *planet* comes from a Greek word meaning "wanderer." The word originally referred to what people believed were five bright stars that seemed to wander among all the rest of the stars in the sky. Other than this strange movement and their brightness, the five stars seemed to be no different from the thousands of others that filled the night sky. The Romans named these five stars Mercury, Venus, Mars, Jupiter, and Saturn, but cultures all over the world were aware of them and often ascribed special religious or mythological significance to them. According to many interpretations, two planets—Venus and Saturn—are mentioned in the Christian Bible.

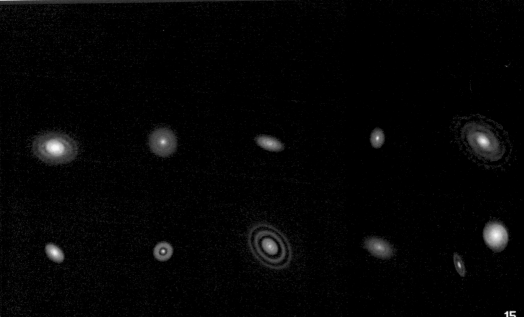

It is uncertain who actually invented the telescope. The first patent was filed in the Netherlands in 1608 by lens maker Hans Lippershey, but the instrument might have already been around for many years. The first telescopes were just a pair of ordinary glass lenses set at either end of a wooden tube, but they made distant objects appear closer. The Dutch were immediately excited by its potential use to navigators and the military. But then Italian scientist Galileo Galilei did something with the telescope that no one else had done before: he turned it toward the night sky. During the winter nights of 1610, he not only discovered that the moon was covered with mountains, valleys, and craters, but he also saw that Jupiter, Saturn, and Venus were not stars at all but worlds like Earth. He saw that Venus had phases just like the moon and Jupiter was circled by four tiny moons of its own.

The ancient Aztecs recorded the movements of the planets—especially Venus, which they associated with war. Astronomers in ancient India included the sun and the moon among the planets, all of which were considered forms of their gods. For hundreds of years, no one bothered to look for new planets because no one believed there *could* be any.

That all changed in 1781 when German-British astronomer William Herschel noticed something unusual while doing a routine star survey with a giant telescope he had constructed. He had spotted, among the familiar stars he was observing, a tiny greenish dot not much larger than a pinhead. He had never noticed it before. Over the next few nights, he noticed that it seemed to be slowly moving. He thought at first that he had discovered a new comet. Most comets have highly elliptical orbits that are very different from the nearly circular orbits of most of the planets. But when Herschel worked out the orbit of the mysterious object, he found that the little green

William Herschel learned how to build his own telescopes from a local mirror maker. Large telescopes such as the one depicted here use multiple curved mirrors to focus the light from distant, dim objects into the observer's eye.

dot had an orbit that was nearly circular. He realized that it must be a new planet. Herschel's news took the world by storm, and there was a rush to name the new planet. The names Georgium Sidus (for England's King George III) and Herschel were two popular candidates, but Uranus, after the Greek god of the sky, won out.

When astronomers checked the records left by earlier observers, it turned out that Uranus had been observed at least a dozen times between 1690 and 1781, but because it was so dim and moved so slowly, no one had recognized it for what it was. And when these old records were carefully reviewed, the astronomers noticed something strange.

By using those records of the earlier observations, it was possible to compute a very accurate orbit for Uranus. The problem was that in 1781 and after that, Uranus wasn't in the position that the numbers predicted it should be. In some years it seemed to lag behind its predicted position, and

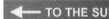

TERRESTRIAL PLANETS   **ASTEROIDS**   **GAS GIANTS**

← TO THE SUN

ROCKS & METALS DOMINATE

in other years it was ahead. Was there some mistake in the calculations? In 1834 the Reverend T. J. Hussey made a startling suggestion: the fault wasn't in the mathematics at all. What if there were yet one more unknown planet orbiting beyond Uranus? Its gravitational pull upon Uranus might account for the discrepancies. When the unknown planet was ahead of Uranus in its orbit, its gravity would tug on Uranus, making it move a little faster. Whenever the unknown planet was behind Uranus, its gravity would cause Uranus to slow down slightly.

Hussey suggested that it might be possible to predict the location of this mysterious planet by working backward from its effect on Uranus. A brilliant young mathematics student at Cambridge University, John Couch Adams, accepted the challenge, and by 1845, he had worked out just where he thought the new planet ought to be. He sent his results to George Airy, the Astronomer Royal (the official astronomer of the British royal court). If Airy were to look at a certain place in the sky, Adams wrote, at the right time Airy would find a new planet. But Airy was skeptical and did nothing with Adams's calculations. Meanwhile, the French astronomer Urbain Le Verrier had made his own calculations and published the results in 1846. He had used the same reasoning as Adams had, and his predicted location for the new planet was almost exactly the same. When Airy saw this announcement, he asked two astronomers—James Challis and William Lassell—to search for the planet. If there *was* a new planet, he wanted it to be discovered by British astronomers, not French ones.

Challis recorded an observation of what he thought was the new planet on August 4, 1846, and again on August 12 but failed to check his

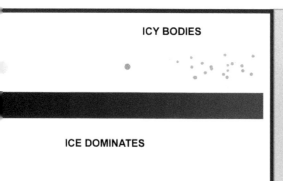

ICY BODIES

ICE DOMINATES

This diagram shows the arrangement of objects in the solar system, with the terrestrial planets closer to the sun and the gaseous and icy objects farther from the sun. By looking at the composition of the various objects in the solar system, scientists can make inferences about what the protoplanetary disk may have been made of.

observations to make sure the object wasn't a known star or a comet. Before he had a chance to do this, Johann Gottfried Galle and Heinrich Louis d'Arrest of the Berlin Observatory in Germany, using Le Verrier's figures, found and identified the new planet. It was named Neptune, after the Roman god of the sea.

The discovery of Neptune was a triumph for mathematics and scientific reasoning. Astronomers became confident that what had been done once might be done again. "There is no reason," Le Verrier wrote, "to believe that this planet is the last one in the solar system. This success allows us to hope that after thirty or forty years of observation of the new planet, we should be able to use it in its turn for discovering the planet next in order of distance from the sun."

That took eighty-four years.

CHAPTER 2

The Planet Hunt

The farther a planet is from its star, the more slowly it moves in its orbit. Mercury, the planet closest to the sun, orbits in a mere 88 days. Earth, almost three times farther away, takes 365 days (1 year), while Neptune requires 165 *years*, twice as long as Uranus. Not even one Neptunian year has gone by since its discovery. It would take a very long time for Neptune to move far enough in its orbit for astronomers to be able to detect the disturbances that would indicate the presence of another, more distant planet, especially if that planet was small with a proportionally small gravitational pull.

The search for a planet beyond Neptune was abandoned until the early 1900s when a wealthy amateur astronomer from Boston accepted the challenge. Percival Lowell was a brilliant student from a distinguished family (his brother became president of Harvard, and his sister won the Pulitzer Prize for poetry in 1926). He graduated from Harvard in 1876 with honors in mathematics. Then he managed his family's cotton mills and electric companies but soon became bored.

He had long been interested in astronomy and decided to make it his career. He had a special fascination with the planet Mars and, being rich, was able to build his own observatory in 1894 near Flagstaff, Arizona, on a 7,000-foot (2,134 m) elevation he called Mars Hill. In 1908 astronomer Earl C. Slipher, who shared Lowell's interest in Mars, joined him. Lowell Observatory is still operating, specializing in observing the planets.

Lowell wondered if he could find a new planet by ignoring Neptune and concentrating on Uranus. His calculations of Uranus's orbit would require much more precision than those of Adams or Le Verrier. Any disturbances in the orbit of Uranus, however slight, that Neptune could not account for must, he reasoned, be caused by yet another planet. His calculations took him years, but in 1905 Lowell announced that he had determined the orbit of what he called Planet X. He even published a description of the planet. It would, he said, be a small world 4 billion miles (6.4 billion km) from the sun—more than forty times farther than Earth—taking 282 years to make a single orbit. Something that small and that far away would of course be extremely faint. Lowell had a significant advantage over his predecessors of the previous century: the camera.

Adams and Le Verrier had to do their observations visually, meticulously marking by hand on their star charts the location of the dim objects they were observing. No matter how carefully they did this or how accurate their charts were, it was an extraordinarily difficult and laborious task, and one always liable to mistakes. The observations were only as good as their eyes, their charts, and the steadiness of their hands. But with a camera, Lowell and his staff could take a picture each night of the part of the sky that interested them. And the camera's film can be exposed for a long time. This allows light to build up on the film, making extremely dim objects much more visible than they would be to the eye.

By carefully comparing photographs, Lowell would be able to tell if one of the tens of thousands of points of light had moved over time. It was still a daunting task, because even on a large photographic plate, Planet X would still be small and dim. Hundreds of faint stars appeared in each photo, and Lowell had to compare each star to those in other photos to see if one had moved. By the time Lowell died in 1916, he had not discovered his mysterious Planet X.

Interest in searching for a new planet died with Lowell. His observatory, though, continued to operate, and thirteen years later, in 1929, a new instrument was installed: a combination telescope-camera that could detect objects many times dimmer than could the instrument Lowell had been using. Interest in finding Planet X was renewed. V. M. Slipher, the director of the observatory and brother of Earl Slipher, assigned the task of searching for it to a twenty-three-year-old astronomer from Kansas, Clyde Tombaugh.

Even with the new, improved equipment, Tombaugh's task was very difficult. The camera would take a photograph of a small portion of the

A blink comparator allows an astronomer to switch quickly back and forth between two images of the same spot in the sky. These instruments are no longer used in contemporary astronomy, as astronomers can use computer programs to detect differences in the brightness or position of objects between multiple digital images.

night sky. Two or three days later, it took a second photograph of exactly the same location. Tombaugh then placed the two photographs into a blink comparator. This device allowed him to compare the two photographs by flipping the images back and forth quickly, like the frames in an animated cartoon. Any fixed points of light, such as stars, would appear unchanged. But if something had moved in between the two exposures, it would appear to "jump" back and forth as the plates were compared. But it was not as easy as it might at first sound. The camera produced hundreds of plates for Tombaugh to examine. Each image could contain up to four hundred thousand tiny points of light. Any motion would be extremely small and difficult to detect, even with the blink comparator. It took Tombaugh a year.

On February 18, 1930, Tombaugh was examining a pair of plates taken of a region in the constellation Gemini when he saw a tiny speck of light jump as he compared the images. It was only a slight difference. The speck moved scarcely one-seventh of an inch (0.36 cm). He compared the plate to others taken of the same area to make certain that he wasn't being fooled by a flaw in the photo. The same faint spot was in all of them. It took some time to officially confirm Tombaugh's observation, but on March 13, 1930—the anniversary of Lowell's death—Slipher announced that Lowell's Planet X had been discovered.

Everyone wanted to name the new planet, the first one to be discovered in nearly a century. The Lowell Observatory—which had the right to decide on the name—chose to call the new planet Pluto, a name suggested by Venetia Burney, an eleven-year-old schoolgirl in England. Pluto was the Roman god of the underworld, and the first two letters of the name—which also form the symbol for the planet—were Percival Lowell's initials.

AT THE EDGE OF THE FRONTIER

Astronomers knew little about Pluto until 2015. The tiny world is so small and distant that it appeared as only a tiny dot in the best Earth-based telescopes, and even photographs taken by the powerful Hubble Space Telescope revealed only a blurry pink object with vague dark and light patches. All anyone really knew about Pluto was not much more than what Lowell had predicted: it was a small world that took more than two hundred years to circle the sun.

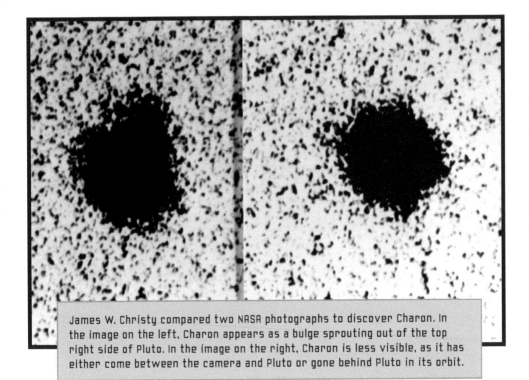

James W. Christy compared two NASA photographs to discover Charon. In the image on the left, Charon appears as a bulge sprouting out of the top right side of Pluto. In the image on the right, Charon is less visible, as it has either come between the camera and Pluto or gone behind Pluto in its orbit.

Most astronomers expected Pluto to be little more than a small, barren ball of rock and ice, and if it had an atmosphere, it would spend most of Pluto's 248-year-long orbit frozen onto the surface. Almost the only new thing that had been discovered about Pluto since 1931 was that it had a moon.

In 1978 Naval Observatory astronomer James W. Christy had been examining a telescopic photo of Pluto. As with most photos of Pluto at the time, it was just a small, blurry blob. But where the blob was usually circular, this time, it seemed to have grown a lump on one side. A later observation showed that the lump had changed position. Christy realized that the "lump" must be a large moon emerging from behind the little planet. Astronomers named this moon Charon.

Charon—named for the boat operator who ferried souls to the underworld in Greek mythology—turned out to be even more of a surprise than expected. It is the largest moon in the solar system compared to the planet it orbits. Earth's moon is about one-fourth the size of Earth, but Charon is almost one-third the size of Pluto.

A DOUBLE WORLD

The Pluto-Charon system is not a moon orbiting a planet but an example of a binary planet. When a moon orbits a planet, it orbits a point, called the center of mass, or barycenter, that lies between the centers of the planet and the moon. For planets with very small moons, such as Mars, the barycenter is almost exactly at the center of the planet. But for larger moons, the center of mass is somewhere between the center of the planet and the moon. The barycenter for Earth and its moon is a point about 1,056 miles (1,700 km) below the surface of Earth, or nearly 2,902 miles (4,671 km) from the center of the planet.

With one exception, all the solar system's moons orbit barycenters that lie beneath the surface of their planet, just as the barycenter for Earth and its moon does. The exception is the Pluto-Charon system. The barycenter lies above the surface of Pluto. Instead of a planet-moon system, astronomers consider Pluto and Charon to be a binary planet, the only one known to exist in our solar system.

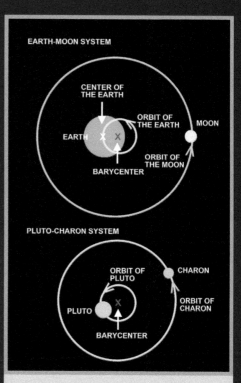

The red Xs in this diagram represent the barycenters for the Earth-moon system (*top*) and the Pluto-Charon system (*bottom*). Both objects in each system orbit around their respective barycenter, but for the Pluto-Charon system, the barycenter lies in the space between the two objects, while for the Earth-moon system, it lies beneath Earth's surface.

But what this moon and Pluto might be like—besides dark and cold—were still mysteries.

THE FIRST MISSION TO PLUTO

Understanding what Pluto is like can tell scientists a lot about what other Kuiper belt objects might be like since Pluto is the innermost member and one of the larger members of the Kuiper belt.

In the 1960s, scientists had realized that the outer planets would soon line up in such a way that a space probe might be able to fly by several different planets in one mission. They pointed out that just two missions could make flybys of Jupiter, Saturn, Uranus, Neptune, and Pluto. But as plans were underway, the US government organization in charge of the mission, the National Aeronautics and Space Administration (NASA), decided that including such a distant world as Pluto would cost too much, so the deep-space probe missions, Voyager 1 and Voyager 2, visited only the four nearer worlds.

At the time, many scientists thought Pluto was little more than a giant ball of ice, an object of little scientific interest. Plans for exploring the distant world slowly faded away.

Throughout the next several decades, probes, orbiters, landers, and rovers explored Mercury, Venus, Mars, Jupiter, Saturn, Uranus, and Neptune. Spacecraft observed some of the solar system's larger moons up close, and a probe had even landed on one of them: Titan, Saturn's largest moon. But one world was still a mystery—little Pluto. Astronomers had long hoped for a mission to the distant, little world, but it would be very expensive. The farthest planet visited had been Neptune, which is thirty times farther away from the sun than Earth is. But that is only 75 percent of the distance to Pluto. It would be a long way to go to visit a world no one knew anything about. NASA was reluctant to invest the billions of dollars such a mission would cost. But the discovery of Charon and evidence for a thin atmosphere on Pluto had revived interest in the little planet, and astronomers were eager to learn more about it.

For a spacecraft to reach such a distant target as Pluto, and do so economically, it needs to get a boost by flying past Jupiter. The giant planet's strong gravity creates a kind of slingshot effect on a passing spacecraft, flinging it away at a greater speed than it had when approaching.

The problem is that for the slingshot effect to work, Jupiter and Pluto have to line up in exactly the right way. That happens only about every twelve years. Scientists missed their chance in 1979 and again in 1991. Seeing an opportunity for 2007, astronomers throughout the 1990s proposed several different missions. NASA dismissed most of these as too expensive or too difficult.

In 2001 astronomer Alan Stern proposed a mission he dubbed New Horizons. A spacecraft would fly past not only Pluto and Charon but several Kuiper belt objects as well. This time, NASA approved a budget for the mission, and by 2005, the New Horizons spacecraft had been assembled and tested.

Meanwhile, interest in Pluto was increasing at a rapid pace. For several years, the planet had been passing through its nearest approach to the sun, causing its atmosphere to warm up and expand. Being able to determine the composition of Pluto's atmosphere would teach astronomers much about the history of the planet as well as reveal conditions on its surface. The Hubble Space Telescope had also discovered two new moons, Nix and Hydra, in 2005.

The New Horizons spacecraft was launched on January 19, 2006. Traveling at more than 36,000 miles (58,000 km) per hour, it was the fastest spacecraft ever launched from Earth. The day it was launched, it passed Earth's moon—a journey that had taken Apollo astronauts three days. Pluto is so distant, though, that even as fast as New Horizons was traveling, it took nine years for the spacecraft to reach the planet. It swung past Jupiter in February 2007 to get a boost from the giant planet's gravity. Six months before arriving at Pluto, New Horizons started studying the planet and its moons. It made its closest approach to Pluto on July 14, 2015, when it zipped past the planet at 36,400 miles (58,580 km) per hour. Despite how quickly it was moving, the spacecraft was still able to send nearly six thousand images of Pluto and its moons back to Earth.

The New Horizons probe was covered in gold-colored foil, which insulates the electronic equipment and keeps it warm enough to operate.

LEARNING ABOUT PLUTO

Scientists and science fiction authors had long been intrigued by Pluto. The first science fiction story set on Pluto was published in 1931, only a year after its discovery, and there have been dozens since. Even John Jacob Astor—one of the wealthiest men in America and founder of the famous Waldorf Astoria hotel—described a very Pluto-like world in a novel he wrote in 1894. Before the arrival of New Horizons, little was known about the new world, other than that it would have to be dark and very cold. No one was even certain how large Pluto was. In the photos taken by the Lowell Observatory, Pluto was just a speck of light. If it were made of dark rock, like an asteroid, it would have to be very large to reflect enough light to show up in a photo. But if it were icy, it would be bright and shiny and could instead be very small.

In spite of what scientists and science fiction writers imagined, Pluto turned out to be a much more complex and much more interesting world than anyone expected. There was not only an icy landscape with craters but also mountains, valleys, plains, and glaciers—and perhaps even cryovolcanoes: volcanoes that erupt cold gases and frozen liquids instead of molten rock.

With a diameter of just 1,430 miles (2,302 km), Pluto is about one-sixth the size of Earth or about two-thirds the size of our moon. Pluto orbits, on average, about 39 AU from the sun. The sun would appear 39 times smaller in Pluto's sky and provide 1,521 times less light and heat. That would still be about 300 times brighter than the light from our full moon. Pluto is so far away that the sun's light takes five and a half hours to reach it.

Scientists had already known that Pluto would have to be very cold because of its vast distance from the sun. But New Horizons let them know exactly how cold the little planet really is. With so little heat arriving from the sun, the temperature on Pluto can be as cold as –375°F to –400°F (–226°C to –240°C). This is so cold that substances such as nitrogen, carbon monoxide, and methane, which are normally found as gases on warm and habitable Earth, are frozen as solid as rock on the surface of Pluto. Water also exists on the surface, but on Pluto, it's just another form of rock.

Because Pluto's gravity is so weak, its atmosphere can reach incredible altitudes, extending as far as 1,100 miles (1,770 km) above Pluto's surface.

Long before the New Horizons flyby of Pluto, scientists had determined that the little planet had an atmosphere by watching stars as they passed behind Pluto. If there were no atmosphere, they would have blinked off as they passed behind the planet and then blinked back on when they reemerged. Instead, the stars dimmed gradually before blinking off, providing evidence of an atmosphere.

Pluto's atmosphere is almost entirely nitrogen, the main component of

Earth's atmosphere. The atmosphere is very thin, though. At the planet's surface, the air pressure is one hundred thousand times less than that on the surface of our planet.

This atmosphere is not constant but varies depending on the distance of Pluto from the sun. When the planet is nearest the sun, there is just enough warmth to cause some of the ice on Pluto's surface to sublimate, or turn directly into a gas. If you have ever seen a block of dry ice—or frozen carbon dioxide—slowly disappear, you have witnessed sublimation.

Aboard New Horizons is a special camera that applies a process to raw color images to bring the photographs closer to "true color," the colors that the human eye would perceive.

The dry ice turns into gas without first passing through a liquid phase. But enough methane is in Pluto's atmosphere to create a measurable greenhouse effect, a warming process that occurs when a planet's atmosphere traps heat arriving from the sun. The increased warmth contributes to the sublimation of Pluto's surface ice.

Pluto came closest to the sun in 1989 and has been moving farther away ever since. As it moves farther away from the sun, it gets colder. Its atmosphere is beginning to freeze. Eventually almost all its atmosphere will become ice on its surface, leaving Pluto with virtually no atmosphere.

As incredibly thin as Pluto's atmosphere is, it is still enough to support a low haze and even light clouds. Even winds blow between the coldest and warmest sides of the planet, occurring at about 60 miles (97 km) above the surface and traveling at speeds ranging from 23 to 225 miles (37 to 362 km) per hour. Winds nearer the surface blow at a gentler 20 miles (32 km) per hour.

A VISIT TO PLUTO

For a small world, Pluto has amazingly varied terrain. Scientists looking at the hundreds of photos returned by New Horizons found that the little planet was a lot more than the barren ball of ice they had thought it would be. There are mountains, plains, valleys, craters, and even glaciers. There are signs of erosion and what appear to be ancient cryovolcanoes, hinting at a much more active world than anyone had expected.

One thing about Pluto that's hard to miss is its color. It's *pink*. The color is from the methane in the ice that covers the surface. When cosmic rays, or fast-moving charged particles, and infrared radiation from the sun strike the surface, they interact with the methane in chemical reactions to create hydrocarbon molecules. These large, organic molecules are often red. As a result of this same effect, Saturn's moon, Titan, is red and orange. Many Kuiper belt objects are red.

Because of how close it is to the rubble-filled Kuiper belt, Pluto has hundreds of craters. Some are as large as 162 miles (260 km) in diameter. But scientists noticed something odd about the craters on both Pluto and its moon Charon: none of them were very small. The craters on Mars and our moon, for instance, come in every size, from tiny pits a few feet wide to holes hundreds of miles across. Most of the craters on Pluto and Charon were relatively large, suggesting that the Kuiper belt is composed mostly of fairly large objects.

Most of Pluto's surface is covered by large, flat, icy plains. These plains have almost no craters, suggesting that they are relatively young—that is, some process is erasing new features such as craters.

If ice covers a broad plain, such as the ice sheet covering Antarctica, the ice will tend to flatten out. Any irregular features, such as hills or depressions, will eventually disappear. It is like taking a spoonful from a bowl of oatmeal. The hole the spoon makes will remain for a short while, but eventually the surface of the oatmeal will flatten out. Scientists believe it is this effect that causes the craters on Pluto to disappear, though it may take decades or even centuries.

The largest of Pluto's flat plains is the giant heart-shaped one that is the planet's most distinguishing feature. Named Sputnik Planitia, it is a nitrogen glacier that covers 347,000 square miles (900,000 sq. km), about the size of Texas and Oklahoma combined. It is 652 miles (1,050 km) wide and at least 2.5 miles (4 km) thick. It originally formed near the north pole of Pluto when a large Kuiper belt object collided with Pluto four billion years ago. The impact created a deep basin in which Sputnik Planitia formed.

This huge mass of ice caused Pluto to wobble, like an unbalanced top. The glacier's weight, combined with the tidal pull of Charon, has caused Pluto's axis to tip 57 degrees, so that it seems to rotate on its side.

Something even more massive added to the weight of Sputnik Planitia's ice. Scientists suspect that an ocean of liquid water is trapped under the nitrogen ice. The ancient impact that created the basin removed a lot of Pluto's crust, leaving only a thin shell. Water could have leaked through this, and the nitrogen glacier eventually formed on top.

INSIDE PLUTO

In looking at photos of Pluto's surface, scientists were surprised to see that many craters showed signs of erosion or were partially filled in, suggesting that Pluto might be geologically active. They also noticed several mountains with craters on their summits, what looked exactly like volcanoes on Earth:

cryovolcanoes. The material from these volcanoes would eventually fill in the oldest craters. Near the cryovolcanoes are plains where cryolava—cold liquids—flowed over the surface and then froze. The Viking Terra region is covered with cracks and shallow, flat-floored valleys that formed as the cryolava cooled. Another region, Virgil Fossae, is covered with frozen cryolava that is rich in ammonia compounds that have colored the ice red.

Another clue that Pluto has an active interior is that Pluto's plains are broken up into polygon-shaped plates. Instead of being flat, unbroken expanses, they look as though they are made of irregularly shaped tiles or flagstones. Convection may have caused these polygonal shapes. When you were cooking that pot of oatmeal, you may have noticed something similar as the hot oatmeal rose, making patterns in the surface.

The resurfaced terrain and the presence of cryovolcanoes together provide evidence for one of the biggest surprises to come from New Horizons: Pluto may have a warm interior. And if it does, it may have lakes of liquid water deep beneath its icy crust. The presence of water would explain many of the mysterious features on Pluto, such as the regions that seem to be very young. Similar features are on Europa and Enceladus, moons of Jupiter and Saturn, respectively. Liquid water flowing from deep beneath the moons' surfaces or ejected by geysers and cryovolcanoes creates an icy coating, filling craters and covering up older features with a fresh new surface.

If Pluto does have an icy crust floating on top of liquid water, the crust will move around, like ice floes in the Arctic and Antarctic on Earth. The movement of the ice will also renew and alter the surface as the ice cracks and as blocks of ice run into and over one another. Astronomers estimate Pluto's icy crust to be about 110 miles (177 km) thick, with a sea of warm, liquid water about 120 miles (193 km) deep between the crust and the rocky core.

The possible existence of lakes or a sea of warm water beneath Pluto's icy crust leads to an idea that no one would have thought possible a few decades ago: life might exist on Pluto. Pluto's seas would have everything life needs to evolve: liquid water, organic chemicals (such as methane and ammonia), and a source of energy—the warmth leaking from Pluto's core instead of the sunlight that triggered the origins of life on Earth.

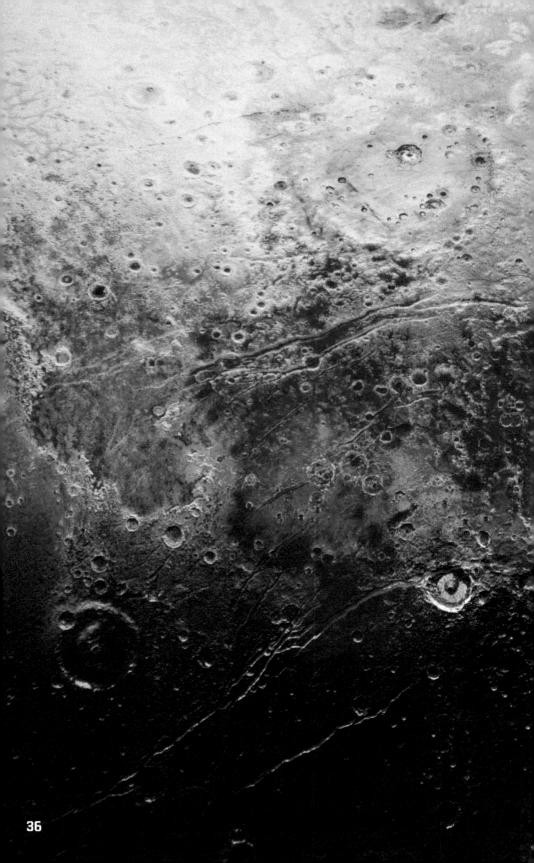

NASA scientists were surprised by the unusual terrain pictured here on Pluto's surface, captured by the New Horizons spacecraft. They believe the ripples and folds are caused by Pluto's active interior.

GLACIERS

Pluto has many glaciers, similar to those on high mountains on Earth or in polar regions like Greenland or Antarctica. Unlike Earth's water ice glaciers, Pluto's glaciers are made of frozen nitrogen. On Earth, glaciers slowly flow downhill. The same thing happens on Pluto. Nitrogen ice flows from the surrounding highlands into Sputnik Planitia, carving valleys along the way. As the nitrogen ice flows over the surface, some of it will melt. Since liquid nitrogen is less dense than solid nitrogen, it will rise to the surface through cracks and fissures. It will flow over the surface of the glacier or perhaps even erupt in geysers.

Pluto's glaciers and icy plains also contain many small, isolated features that could be mountains or icebergs. Water ice is less dense than nitrogen ice, so blocks of water ice, torn away by the slowly moving glacier, might have floated up from deep beneath the surface. Astronomers have seen many of these in Sputnik Planitia.

MOUNTAINS, VALLEYS, AND DUNES

Pluto has regions of rugged mountains that can be as high as 6,500 to 11,000 feet (1,981 to 3,353 m). Instead of rock, like the mountains on Earth, Pluto's mountains are made of giant blocks of ice. On frigid Pluto, ordinary water ice is as hard as rock. With their snowcapped peaks, these mountains would look familiar to anyone who has seen any of the high mountains on Earth. But instead of water ice, the snowy caps of Pluto's mountains are made of frozen methane.

The mountains of Earth's moon have been worn down by billions of years of erosion, so they resemble softly rolling hills. This erosion has been caused by the impacts of micrometeorites, wearing them down like sandpaper, and by thermal erosion, which occurs when changes in temperature causes the surface of a rock to crack and flake away. But Pluto's mountains look very young. They are sharp and rugged-looking and contain very few craters. It takes time for ice to erode or flow to eventually smooth out features like mountains and craters. So, scientists believe that while the rest of Pluto's landscape is billions of years old, its mountains may have been created relatively recently, perhaps only one hundred

In addition to giant Charon, discovered in 1978, Pluto has at least four other moons. All of them are very tiny. Nix and Hydra were both discovered in 2005 by the Hubble Space Telescope. Orbiting between them is Kerberos, discovered in 2011. Astronomers found Styx in 2012. The largest of these four are Nix and Hydra, at 31 miles (50 km) and 32 miles (51 km) wide, respectively. The smallest is Styx at only 6 miles (10 km) wide. None of these moons are round like Earth's moon. Instead, they resemble lumpy rocks or potatoes.

million years ago. And scientists believe that Pluto's mountains may still be growing.

Pluto has valleys and fissures as long as 370 miles (595 km) and as deep as 2.5 miles (4 km). They could have formed from underground water. As water freezes, it usually expands. Under certain conditions of temperature and pressure, freezing water will also contract. The expansion and contraction of underground water could have created cracks in Pluto's surface. Most of this would have occurred long ago when the surface ice was relatively thin. Today, any underground oceans are buried beneath a layer of ice 200 miles (320 km) thick.

A surprising discovery was that Pluto has dunes. They look almost like the sand dunes in the American Southwest or in deserts like the Sahara. A large area near the edge of Sputnik Planitia is covered with them. Instead of sand, Pluto's dunes are made of tiny particles of ice blown by the planet's gentle winds.

CHAPTER 3

The Mystery of the Kuiper Belt

Sometimes a strange visitor appears in the night sky. It looks like a star with a long glowing trail behind it. The ancient Greeks called these visitors *comets* from a word meaning "hairy" or "bearded," because they thought that comet tails looked like long, flowing hair.

In 1705 British astronomer Edmond Halley discovered that comets travel in long, egg-shaped orbits around the sun. By working out the orbits of four famous comets—those seen in 1456, 1531, 1607, and 1682—he found they all had exactly the same orbit. They also all appeared exactly seventy-five years apart. He realized that what seemed to be four different comets was just one that appeared repeatedly as it swung around the sun in its orbit. He predicted that the comet would appear again in 1758, and on Christmas night of that year, the comet returned. It was named Halley's comet in his honor. It last appeared in Earth's skies in 1986 and won't appear again until 2062.

WHERE *ARE* COMETS?

Astronomer Fred Lawrence Whipple famously described comets as being little more than "dirty snowballs." They resemble enormous icebergs composed of a mixture of ice, frozen gases, large rocks, gravel, and dark, carbon-rich material. Most comets are only 0.5 to 12 miles (0.8 to 19 km) wide, much too small to be seen with telescopes. Only when a comet heads toward the sun and warms up can it be seen.

Beyond Jupiter and Saturn, sunlight is much too weak to provide much warmth. The ice in comets orbiting there is frozen as hard as steel. But when a comet comes to within about 3 AU, about twice the distance of Mars, it starts to warm up. Some of its ice, made mostly of water and carbon dioxide, begins to sublimate. The closer the comet gets to the sun, the more its ice turns to gas. The gas forms a kind of atmosphere around the comet called a coma. It glows brightly in sunlight. Astronomers see this, not the tiny, dark nucleus of ice and rock in the center.

Comet NEOWISE came close to Earth in early 2020, brightening enough to be visible to the naked eye for many weeks. At its closest, it was only 0.7 AU away from Earth.

As the comet gets closer to the sun, the coma gets larger and larger. It eventually becomes hundreds and even thousands of miles wide. The solar wind, a stream of atomic particles emitted by the sun, will cause the gases to stretch out into a long, flowing tail. The tail of a comet is like a flag blowing in a breeze. It may be millions or even tens of millions of miles long.

WHERE DO COMETS COME FROM?

A question that puzzled many scientists was, Where do comets come from? Since comets don't last forever, scientists reasoned that there must be a kind of reservoir in deep space from which new comets were constantly replacing old ones.

Shortly after the discovery of Pluto, several scientists began wondering what lay farther out in the solar system. Some of the material left over from the formation of the solar system is found in the asteroid belt—the region between the orbits of Mars and Jupiter. Scientists believed that the rest of the leftovers wound up far beyond Pluto. Because most of the heavier elements fell in close to the sun, the debris beyond Pluto would be composed mostly of ice. They would range in size from snowballs to objects tens or even hundreds of miles wide. Occasionally, one of these mountains of ice would fall toward the sun. When it did, it may become a comet.

This idea of a distant reservoir of comets came from studying short-period comets. They have two special characteristics. First, it takes them only two hundred years or less to orbit the sun, so their aphelion, the farthest point their orbits take them, lies just beyond Pluto. Second, they orbit the sun more or less in the same plane as the planets.

These two facts were strong evidence for the existence of a swarm of icy bodies circling the solar system not far outside the orbit of Pluto. Although a number of different scientists proposed the idea at around the same time, the region was eventually named after just one: Dutch astronomer Gerard Kuiper. In 1951 he published the idea that the solar system is surrounded by a vast belt of icy bodies stretching between 30 AU and 50 AU from the sun—from about the orbit of Neptune to just beyond the orbit of Pluto. Kuiper thought that some of these icy bodies might be the source of short-period comets. Some of these bodies might be as large as Pluto. This region was eventually named the Kuiper belt and was believed to be a broad, flat ring, a little like the rings around Saturn.

But no one knew for sure that such a region actually existed. The objects there would be far too small and far away to be detected directly by telescopes, at least not those available at the time.

In 1977 astronomer Charles Kowal discovered the first of an unusual new type of object in the solar system. Chiron was a small, icy, comet-like body orbiting between Saturn and Uranus. Measuring between 123 and 148 miles (198 and 238 km) wide, it was the first of a large number of similar objects, or Centaurs, discovered orbiting between Jupiter and Neptune. They are very much like comets, and as the orbit of each takes it close to the sun, a little bit of its surface will sublimate. Eventually, over hundreds or thousands of years, a Centaur will disappear, like an ice cube on a warm day. But since so many were still in the solar system, new ones must be coming from somewhere. One possibility was a cloud of icy bodies beyond Pluto, just as Kuiper and other astronomers had predicted.

DENIZENS OF THE KUIPER BELT

In the late 1980s, astronomers David Jewitt and Jane Luu began looking to see if Kuiper belt objects might actually exist. They succeeded in 1992 with the discovery of 1992 QB1, later named Albion. About 75 miles (121 km) wide, it was a very small object, but it was the first to be discovered orbiting in the space beyond Pluto.

Astronomers Michael Brown and Chad Trujillo discovered Quaoar in 2002. At 690 miles (1,110 km) in diameter, it was half the size of Pluto. It was the first really large Kuiper belt object found. Sedna, discovered in 2003, is 309 miles (497 km) in diameter, while Eris, announced as a discovery in 2005, is nearly as large as Pluto. To be spherical, a body made of ice needs a minimum diameter of about 250 miles (400 km). This meant that some Kuiper belt objects were large enough to qualify as planets, according to the Geophysical Planet Definition.

Most large Kuiper belt objects that have been observed are spherical. For an object to be seen at such great distances, it needs to be fairly large.

WHAT *IS* A PLANET?

The solar system is composed of many different kinds of objects such as planets, asteroids, Kuiper belt objects, satellites, and comets. Scientists have been arguing for decades about where to draw the lines separating them.

One of the roles of the International Astronomical Union (IAU) is to decide on definitions. At a meeting in 2006, the IAU voted on a new definition of the word *planet*. The new definition demoted Pluto from *planet* to *dwarf planet*. This decision upset many planetary astronomers who felt that they had not been given enough say in the vote and that only planetary astronomers should be able to define the terms they use in their work. Like the field of medicine, astronomy has many specialties. Just as dentists, surgeons, and heart specialists are in medicine, some astronomers specialize in astrophysics, cosmology, and planets. The planetary scientists felt that astronomers who are experts in stars and galaxies should not make decisions on behalf of scientists who are experts in planets.

So planetary scientists came up with their own definition. They felt that the Geophysical Planet Definition was more accurate and more useful than the one devised by the IAU. In its simplest form, the definition states that a planet is any body smaller than a star that has never undergone nuclear fusion and has enough gravitation to be round due to hydrostatic equilibrium—or large enough that its shape is determined primarily by gravity. Some of the very largest asteroids, such as Ceres or Pallas, would be considered planets by this definition since they are large enough to be round, while the majority of other asteroids are small and irregularly shaped. While the IAU definition of *planet* remains the official one, the majority of planetary scientists use the Geophysical Planet Definition in their work.

And if an object is large enough, it will probably be spherical. Haumea, discovered in 2004, is an exception. It is 1,430 miles (2,300 km) at its equator, large enough to be round. But instead, it is shaped something like a football.

Its shape is probably due to its fast spin. It rotates once every four hours, the fastest spin of any large object in the solar system. So, while Haumea is over 1,000 miles (1,609 km) long, it is only 619 miles (996 km) wide. This rapid spin was probably created by a large object colliding with Haumea millions of years ago.

But Haumea's shape wasn't the only surprising thing about it. Like the planet Saturn, it has a ring. Unlike Saturn's, its ring is dark, thin, and faint—too faint to be seen directly. Astronomers only realized it was there when Haumea passed in front of a distant star. The star dimmed briefly, and then when it dimmed a second time, scientists realized that a dark ring must be surrounding Haumea. And Haumea also has two tiny moons, Hiʻiaka and Namaka.

Many asteroids and Kuiper belt objects look like miniature worlds. Below are asteroids (Ceres, Pallas, Vesta, and Hygeia) and Kuiper belt objects (Eris, Makemake, Haumea, Sedna, Quaoar, Varuna, Pluto, and Charon) shown to scale with Earth and its moon.

Eris

Makemake

Haumea

Sedna

Quaoar   Varuna

Moon

Ceres   Pallas   Vesta Hygeia

Earth

Charon

Pluto

In 2005 astronomers announced the discovery of Eris. It is 1,445 miles (2,326 km) wide—just slightly smaller than Pluto—making it the second-largest known Kuiper belt object. Eris also has a tiny moon, Dysnomia. Eris is denser than Pluto, suggesting that it is composed of more rock and less ice than Pluto. Eris orbits the sun at about 68 AU, much farther away than Pluto. It takes Eris 577 years to orbit the sun just once, but its twenty-five-hour day is not much different from our own.

Makemake was also discovered in 2005. It is 932 miles (1,500 km) wide, about two-thirds the size of Pluto, making it the third-largest known Kuiper belt object. Like Pluto and many other Kuiper belt objects, it is reddish. Like Eris, Makemake has a tiny moon, MK2. It's about 100 miles (160 km) in diameter and orbits the Kuiper belt object at a distance of about 13,000 miles (20,900 km). Makemake and its moon orbit the sun at a distance not too different from Pluto's, ranging from 53 AU to 38 AU.

By 2022 more than three thousand Kuiper belt objects were discovered. Astronomers believe there are probably hundreds of thousands larger than 60 miles (100 km) wide. All seem to be more or less alike: small, cold, and reddish.

Standing on the surface of a Kuiper belt object, the sun would not appear much larger than the more distant stars in the sky.

WHERE DO THEY GET THEIR NAMES?

The IAU has set up rules for naming objects in the solar system. Place-names on Pluto follow the lead set by eleven-year-old Venetia Burney when she suggested that the planet be named after the Greek god of the underworld. The IAU decided that features on Pluto should be named after mythological gods, goddesses, and explorers of the underworld. Pluto's moons are also named for characters and creatures from Greek and Roman mythology. Charon, for instance, runs the ferry that carries souls across the river Styx, while Kerberos and Hydra were the names of two monsters from the underworld.

Kuiper belt objects are named for a deity or figure related to creation. For example, Makemake is the Rapa Nui creator of humanity and god of fertility, and Haumea is the Hawaiian goddess of fertility and childbirth.

FARTHER AND FARTHER AWAY . . .

Another strange little world was discovered in 2015. Officially named 2015 TG387, it was dubbed the Goblin by its discoverers. The IAU later gave it the official name Leleākūhonua, "where heaven meets the earth" in Hawaiian. It is only about 190 miles (306 km) in diameter, about one-eighth the size of Pluto.

Leleākūhonua is one of just a few objects whose elliptical orbits never take them closer to the sun than Neptune. Just two, 2012 VP113 and Sedna, have a perihelion (the point of an object's orbit closest to the sun) farther out than Leleākūhonua, and its orbit actually takes it far beyond them at its most distant point. Leleākūhonua is about 65 AU at perihelion and is estimated to reach some 2,300 AU from the sun at its farthest, putting it more than twice as far out as Sedna. It takes Leleākūhonua about thirty-two thousand years to orbit the sun once.

In 2018 astronomers found what they thought was an object orbiting the sun at the greatest distance yet known: 120 AU, or 120 times farther from the sun than Earth. It was a small world, about 311 miles (500 km) in diameter, or roughly one-third the size of Pluto. Scientists named it Farout. Only a few months later, though, an even more distant object was discovered. It was 132 AU—12 billion miles (19.3 billion km)—from the sun and was named Farfarout. Slightly smaller than Farout, Farfarout has a very stretched-out, elliptical orbit that carries it as far as 175 AU from the sun to as close as 27 AU, bringing it inside the orbit of Neptune. Astronomers speculate that it might have been a close encounter with Neptune that hurled Farfarout so far into space.

PAYING A VISIT TO A KUIPER BELT OBJECT

Kuiper belt objects are so small and so far away that even in the most powerful telescopes they appear as only tiny dots of light. Astronomers can determine their color and estimate their size, but what they might actually *look* like was a mystery. Most astronomers considered Pluto to be a large Kuiper belt object, but was it a typical example?

After the New Horizons spacecraft made its flyby past Pluto and Charon, astronomers realized they had a chance to get a close-up view of at least one more Kuiper belt object. The New Horizons science team started searching for a possible target. Using the Hubble Space Telescope, they found 2014 MU69, a tiny body right in the path of New Horizons. The spacecraft flew by MU69 on January 1, 2019. It had already been taking photos as it approached. As MU69 got larger and larger in the photos, it became apparent that it was going to be a very strange-looking object. At first, astronomers thought it was actually two small Kuiper belt objects orbiting so close to each other that they nearly touched.

But MU69 turned out to be one of the strangest-looking worlds in the solar system. It looked a little like a snowperson, with a small, round body attached to a larger one shaped like a flattened hamburger bun. It is about 22 miles (35 km) long, 12 miles (20 km) wide, and 6 miles (10 km) thick. Astronomers think that MU69 got its strange shape when two Kuiper belt objects ran into each other and stuck together. The exciting discovery was a kind of fossil remnant from 4.5 billion years ago, when our solar system was first forming. Small objects were colliding and sticking together to

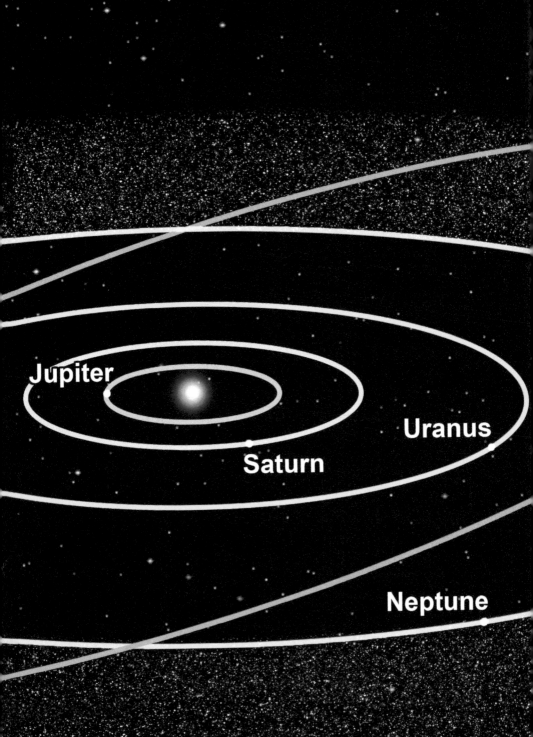

Jupiter

Uranus

Saturn

Neptune

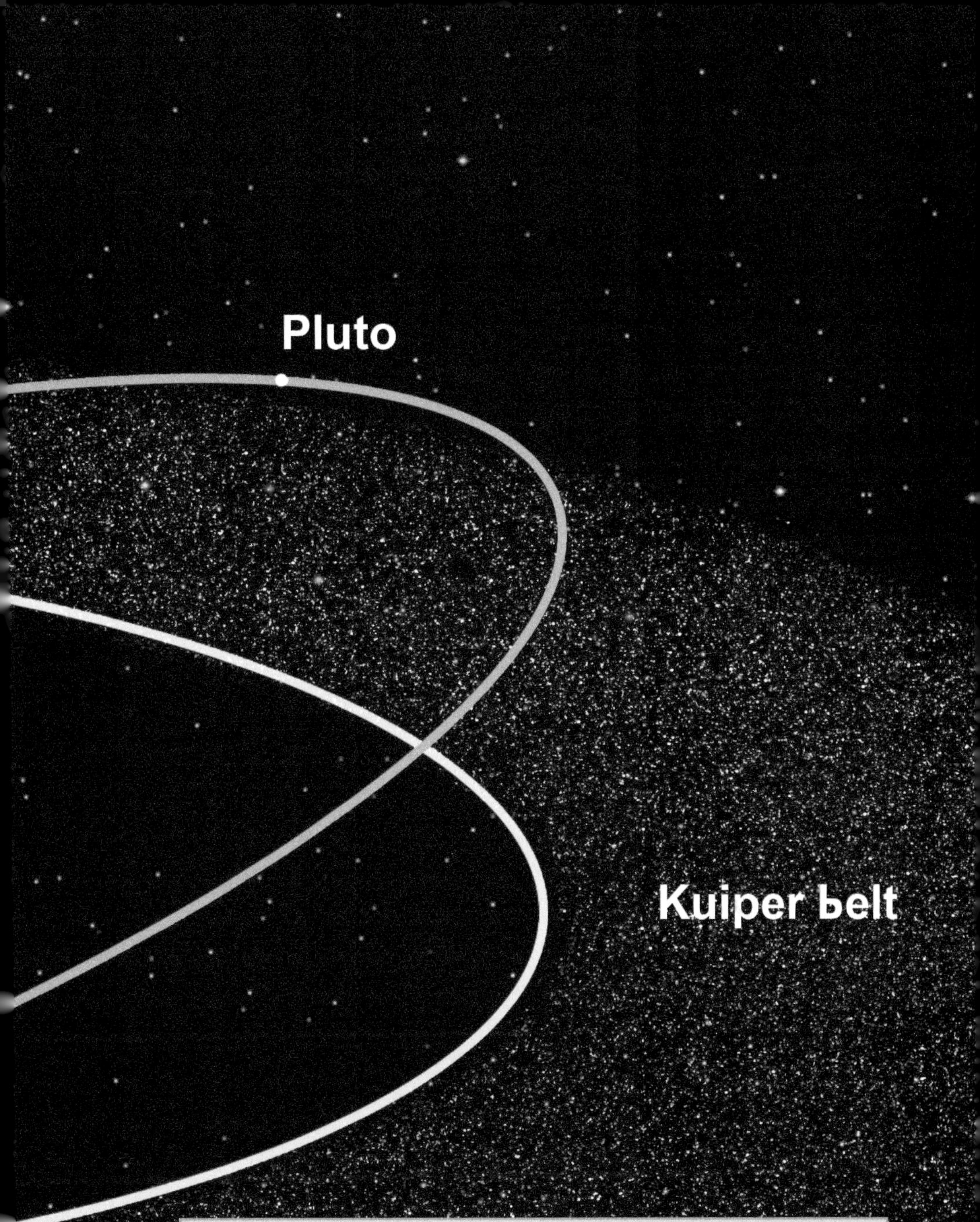

Pluto

Kuiper belt

Pluto's orbit is tilted with respect to the ecliptic plane, an imaginary plane that stretches from the sun's equator through the solar system. Most objects in the inner solar system lie on the plane, providing more evidence that the solar system originated from a thin disk of material. Since the sun's gravity weakens with distance, farther, smaller objects may not necessarily be found on the ecliptic plane.

eventually become planets. MU69 really was a look deep into the past, to the origins of the solar system.

After the New Horizons flyby, MU69 was given the official name Arrokoth, which means "sky" in the Powhatan language.

WHAT *IS* PLUTO?

With so many Pluto-like objects being discovered in the far reaches of the solar system, many astronomers began to wonder if Pluto might belong to the Kuiper belt. The little world was very different from all the outer planets, which are enormous gas giants. Unlike Jupiter, Saturn, Uranus, and Neptune, Pluto is very small—only about two-thirds the size of Earth's moon. The giant planets are made mostly of gas and liquid, but Pluto is made mostly of ice. Most of its orbit lies within the Kuiper belt, a good indication that Pluto may have originated there. It also closely resembles many other objects that have been discovered orbiting in the region beyond Neptune.

Some scientists think that a clue to Pluto's origin lies in its similarities to Neptune's largest moon, Triton. Triton is practically a twin of Pluto. It is nearly the same size, composition, and temperature. It has a strange orbit, going around Neptune in a retrograde direction—opposite to that of Neptune's rotation—unlike every other large moon in the solar system. Astronomers infer from its unusual orbit around its planet that it likely originated elsewhere and was captured by Neptune's gravity after it strayed too close to the gas giant.

Could Pluto also be an object that strayed from the Kuiper belt? Its strange orbit suggests so. All the other planets in the solar system go around the sun in nearly circular orbits. But Pluto's highly elliptical orbit more closely resembles that of a comet. Although its average distance from the sun is 3.6 billion miles (5.8 billion km), forty times farther from the sun than Earth, at its closest approach Pluto actually comes inside the orbit of Neptune. During 1979 to 1999, Neptune was the most distant planet in the solar system. No other planet crosses the orbit of another one.

Pluto's orbit is also tipped in relation to the plane of the solar system, called the ecliptic plane. Most of the other planets lie on or near this plane, which is like a large disk with the planets sitting on it like marbles and the sun in the middle. Pluto's orbit, however, is tipped seventeen degrees with respect to the ecliptic. That means that during its orbit, it first swings high above the plane and then far below.

For these reasons—its small size, its similar neighbors, and its tilted comet-like orbit—many astronomers follow the IAU definition and consider Pluto a dwarf planet. But most planetary astronomers consider Pluto a planet, per the Geophysical Planet Definition. Major astronomical organizations and astronomers continue to debate the definition of *planet* and Pluto's status, but certainly, the little world is full of mysteries and worthy of study regardless of what we call it.

CHAPTER 4

The Dark Frontier: On beyond Pluto

A re there any large planets hidden within the Kuiper belt? At the time that Clyde Tombaugh discovered Pluto, most astronomers thought that even if there were other, more distant worlds, it would be almost impossible to find them. Pluto had been extremely difficult to find, even when Tombaugh knew exactly where to look. Tombaugh flatly declared that "there is no such body as a tenth planet" in our solar system. But many astronomers have begun to wonder.

PLANET X

Tombaugh thought he had solved the problem of the unexplained disturbances in Uranus's and Neptune's orbits when he found Pluto. Pluto is not only the smallest of all the planets, but it is also smaller than at least seven different moons (Earth's moon, Titan, Triton, Io, Europa, Ganymede, and Callisto)! Pluto isn't massive enough to cause the gravitational effects that astronomers had seen in the orbits of Uranus and Neptune. If not Pluto, then what? Several astronomers decided that while Tombaugh had found a new planet, he hadn't found the one that Lowell had predicted. Possibly, the so-called Planet X was still undiscovered.

Planet hunters searched extensively for Planet X in the late 1970s and early 1980s. While most of them used the same "blink" technique that Tombaugh had, technological progress had given them access to much more powerful and sophisticated camera telescopes to obtain their star photos. But despite using improved instruments, no one succeeded. Some still believe Planet X is out there. Why hasn't it been discovered?

Possibly, its orbit is tilted so much to the ecliptic plane that Planet X may lie far above or below the region where most other planets are

The hypothetical Planet X calculated by Brown and Batygin (see p. 57) would take between ten thousand and twenty thousand years to complete one orbit around the sun.

PLANET X AND PSEUDOSCIENCE

Speculation about the possibility of a large planet beyond Pluto has resulted in many people believing in an imaginary planet called Nibiru. Beginning in 1976, author Zecharia Sitchin wrote a series of books theorizing that the solar system contains a mysterious tenth planet. He based this idea on the myths, writings, and artwork of an ancient civilization, the Sumerians, who lived in southern Mesopotamia (where most of modern-day Iraq lies) five thousand years ago. Sitchin claimed that Nibiru has a very unusual orbit that causes it to pass through the solar system every thirty-six hundred years. When this happens, Nibiru (which some call Planet X) can come very near Earth. As it zooms past our planet, it causes cataclysmic events to occur. Sitchin thinks Nibiru's next visit will be in 2087. He believes this close approach to Earth will create worldwide earthquakes, volcanoes, hurricanes, tornadoes, tidal waves, and other disasters.

Supporters of the Nibiru theory say that not only does the planet exist, but it has already been seen. They claim that observatories around the world have seen it and that NASA even has photographs of it! None of this is true, of course.

One reason to dismiss the idea of Nibiru is that any planet of its proposed size would have been easily spotted by astronomers, yet no one has ever seen it. Another is that if a planet the size of Nibiru were anywhere near where believers say it is, its gravitational effects would be easily detected. Finally, while large undiscovered objects may be in the Kuiper belt or beyond, scientists know from the way the solar system formed and from the orbits of known planets that any planet in the outer solar system could not have an orbit that brings it close to Earth.

found—and that is a lot of sky to search if you are not sure of where to look. Another possibility is that Planet X might be so small and dark that it would be lost if seen against the millions of stars in the Milky Way. Its orbit could be so long—perhaps a thousand years or more—that it could take centuries to move into a clear portion of the night sky. Finally, it just might not be there at all.

One reason that many astronomers think a large planet may be hidden deep within the Kuiper belt is that some of the small objects found in the Kuiper belt seem to have strange orbits that seem to be influenced by some large, invisible object. This might be a yet undiscovered planet.

In looking at the orbits of the known Kuiper belt objects, astronomers noticed something unusual. At their respective closest approaches to the sun, the perihelion, their orbits all passed through or near the same point. This suggested that something was disturbing the orbits, something yet to be discovered. Many astronomers thought this mysterious object might be a large, undiscovered planet—something like a dark Neptune, perhaps. They called it Planet Nine, or Planet X.

In 2016 two astronomers, Michael Brown and Konstantin Batygin, calculated the probable orbit Planet X would have. They examined the observations of known Kuiper belt objects whose orbits seemed to be affected by an unknown but massive planet. Since most of the planets (except Pluto) lie in the ecliptic plane, it seemed reasonable to expect Planet X to also be orbiting in the ecliptic plane. The problem was to figure out just *where* along the line of the ecliptic the unknown planet might be. It was a little like knowing the street someone lives on without knowing the address. Adding up all the orbital effects they believed were being caused by an unknown planet, Brown and Batygin determined which region of space along the ecliptic would be the most likely place for Planet X.

The astronomers also made guesses as to its size. It should be, they announced, about 6.2 times the mass of Earth. The planet Uranus, for comparison, is about 15 times the mass of Earth. What kind of planet might it be? It would be about the right size to be a super-Earth. These are planets that are about twice the size of Earth and up to 10 times as massive. They might be rocky, like our planet, or made of ice, gas, or a combination of these. Whatever Planet X may be made of, it would be a very cold world. There are no known super-Earths in our solar system, but astronomers have discovered several orbiting other stars.

The eccentric orbit of Planet X would take it as far as 300 AU from the sun. Brown and Batygin believe that the orbit of Planet X is also probably tipped at about sixteen degrees, so it does not fall exactly along the ecliptic.

Observing Planet X directly depends on a couple of different things. For instance, what *kind* of planet it is. If it is ice-covered, it might be a very bright object, even if it is small. But if it is rocky like Mercury or Mars, it might be darker and harder to see, even if the planet is very large.

Other scientists doubted the existence of Brown and Batygin's planet. The disturbances in the orbits of the Kuiper belt objects are so small, they said, that they could fall well within the margin of error. The margin of error is the difference between a measured value and its true value. When scientists take measurements of very small values, such as the changes in the orbits of very small, faraway objects, the margin of error can be large enough that the measurements become unreliable. Using very precise technology and instruments reduces the margin of error.

Several researchers, including Jakub Scholtz of England's Durham University, had an even more unusual explanation for the data. They proposed that the object might not be a planet. It might be a small black hole—specifically, a primordial black hole. Primordial black holes were created in the first second after the big bang, before even stars and galaxies existed.

Although no one has yet directly observed a primordial black hole, theories about the formation of the universe predict their existence. And if they do exist, they would come in all sizes, from as little as 10^{-7} ounces (10^{-5} g)—about one hundred thousand times less than a paper clip—up to about one hundred thousand times more massive than the sun.

If a small primordial black hole is orbiting in the outer reaches of the solar system, it probably could not be observed directly. Since the black hole would absorb any light that falls onto it, it would be invisible. But scientists at Harvard University have devised a possible way to detect its presence. When matter falls into a black hole, it releases a burst of energy. Scientists can look for the telltale burst when a hapless comet or other body runs into the black hole and is devoured by it.

Black holes are objects that are so massive that not even light can escape their gravitational pull. They are very difficult to observe and can only be detected indirectly through the gravitational effects they have on other objects or through their occasional emissions of energy.

ANOTHER MYSTERY

The existence of the Kuiper belt helped explain the origin of short-period comets. It also helped explain why these comets all entered the solar system more or less on the same plane. But the existence of the Kuiper belt could not be used to explain another type of comet, long-period comets.

Like their name implies, these comets take a long time to circle the sun. Some take many thousands of years to orbit the sun just once. The farthest point in their orbit may be up to 6 trillion miles (9.5 trillion km) from the sun, one hundred thousand times farther than Earth and far beyond the limits of the Kuiper belt. Long-period comets also enter the solar system from every direction. That means their origin couldn't be a relatively flat disk like the Kuiper belt.

In 1950 Dutch astronomer Jan Oort proposed that the solar system might be surrounded by a vast spherical cloud of millions of predominantly icy bodies. The Oort cloud would be far beyond the Kuiper belt, extending from 2,000 to 100,000 AU from the sun, over one-quarter of the distance to the nearest star. This is so large that the two Voyager spacecraft, traveling 1 million miles (1.6 million km) a day, would take thirty thousand years to pass through it. In comparison, the Kuiper belt ranges only from as close as 30 AU to as far as 50 AU from the sun. While there is no direct evidence for the Oort cloud, astronomers have observed similar clouds surrounding other stars, especially those with newly forming planetary systems. Along with the evidence from long-period comets, these observations suggest that things like the Kuiper belt and Oort cloud are a natural part of the planet-star system formation.

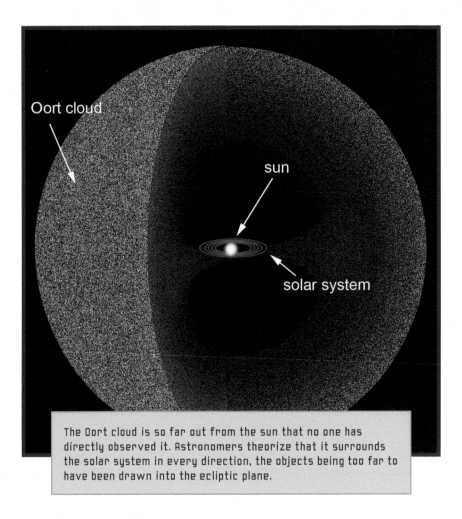

The Oort cloud is so far out from the sun that no one has directly observed it. Astronomers theorize that it surrounds the solar system in every direction, the objects being too far to have been drawn into the ecliptic plane.

During the Late Heavy Bombardment (see p. 63), Earth was hot and molten—not conducive to the formation of life. Scientists theorize that once the bombardment period began to end, the final comets that crashed into Earth brought the water necessary to form life.

Icy bodies drifting around in the Oort cloud are affected by the gravity of nearby stars as well as that of the sun. This tug-of-war can disturb their orbits. Some may escape the Oort cloud and disappear into interstellar space. Others may fall into the inner solar system. As the object gets closer to the sun, its ice will turn to gas. The gas will form a coma around the icy nucleus. The pressure of the solar wind will blow the coma out into a tail millions of miles long. If we are lucky, we might observe it glowing in the night sky as a comet. After a few years, the orbit of the comet takes it around the sun and back toward the Oort cloud. It may be hundreds or thousands of years before it returns to the inner solar system.

Occasionally, one of these comets may pass close to one of the large outer planets, such as Jupiter. The gravity of the planet might disturb the comet's path enough to draw it in closer to the sun, shrinking its orbit. A smaller orbit means the comet will return much more often to Earth's skies. Some scientists believe that this happened with Halley's comet and that it may have originally come from the Oort cloud.

While earlier spacecraft had encountered comets from a distance, such as when the Giotto spacecraft flew past the nucleus of Halley's comet in 1986, the Rosetta spacecraft was the first to rendezvous with a comet and follow it as it cruised through the solar system. Launched by the European Space Agency in 2004, the Rosetta spacecraft arrived at Comet 67P/Churyumov-Gerasimenko in 2014. It even managed to land a probe on the surface of the comet. Philae, the lander, collected data and took hundreds of photos.

What it saw was a strange landscape of rocks, pebbles, and miniature jagged mountains with geyserlike jets of gas erupting from the surface. Although comets are made of up to 80 percent ice, they are surprisingly dark. Every time a comet swings around the sun, more of the bright ice on its surface evaporates, leaving behind the dark organic compounds that comprise the rest of the comet. Over centuries the surface of the comet's nucleus becomes covered in a thick layer of black dust.

The Rosetta orbiter found water vapor in the atmosphere surrounding the comet as well as nitrogen and oxygen. The Philae lander was able to take samples of the comet's surface. It found areas of exposed water ice and even amino acids and phosphorus—both of which are building blocks of life. These discoveries helped support the theory that comets helped start life on Earth by delivering the necessary materials at the end of the Late Heavy Bombardment.

The Rosetta spacecraft took close-up photographs of a jet of icy material emitting from the comet's nucleus. When Philae landed on the comet's surface, it accidentally bounced into a dark crevice where it was unable to collect solar energy. The lander powered down two days after landing, and about a year later, Rosetta concluded its mission by crashing into the comet.

Sometimes a comet will come too close to a planet. In 1992 Comet Shoemaker-Levy 9 came so close to Jupiter that the planet's gravity pulled it apart into a string of fragments each about 1 mile (1.6 km) wide. When the comet returned in 1994, these fragments collided with Jupiter, one after the other, in a series of explosions that were visible in telescopes on Earth.

COMETS AND LIFE ON EARTH

Some scientists believe that comets were important to the beginnings of life on our planet. Earth shortly after its formation was a hostile world. Still hot after accreting from a mass of gas and dust, its surface was mostly molten rock. Continuous impacts from meteors and asteroids kept the surface too warm for the formation of liquid water or the kind of complex carbon molecules necessary for life. This period in Earth's history, the Late Heavy Bombardment, occurred about 3.8 billion years ago. The earliest known forms of life first appeared about 3.5 billion years ago, not long after the bombardment ended. So, life formed very quickly, but how could it appear so fast if the bombardment had destroyed most of the water and organic material? It must have been replaced somehow.

One possibility is comets. Comets are rich in both water and complex carbon-based molecules. When the Rosetta spacecraft explored Comet 67P/Churyumov-Gerasimenko in 2014, it discovered that nearly half of the comet was composed of both simple and complex organic molecules. Rosetta even found the organic molecules known as organic aliphatic compounds. These are thought to be essential building blocks of life.

As soon as the bombardment of meteors, comets, and asteroids eased off and the surface of Earth began to cool, the water and organic molecules brought to our planet by the final comets could remain intact. And with these materials available, life was able to form on Earth.

CHAPTER 5

The Invisible Solar System

To begin exploring the outer limits of the solar system, we need to start back at the sun itself. The effect of its gravity reaches far beyond any planet visible to the most powerful telescopes. Its grip keeps the Kuiper belt and the Oort cloud in place, reaching at least two hundred thousand times farther away than our own world. But the sun creates more than just a powerful gravitational pull.

THE SOLAR WIND

The sun is surrounded by an atmosphere just as Earth is. The sun's atmosphere consists of hot gases and plasma, a superheated substance. Plasma is the fourth state of matter, different from solids, liquids, and gases. It consists of atoms of such a high temperature that the heat has stripped them of one or more of their electrons. Atoms like these are ions. Familiar examples of plasma are bolts of lightning and glowing neon signs.

The sun's corona is usually invisible to the naked eye, but it can be observed during solar eclipses, when the moon crosses in front of the sun and blocks most of the sun's light. Using special light-blocking glasses to protect their eyes from damage, observers on Earth can view the corona.

The outermost visible layer of the sun's atmosphere is the corona. It is made entirely of plasma. It's the part of the sun that creates the streamers visible during a solar eclipse. The plasma becomes so hot that the gravity of the sun can no longer hold it down. Trapped by the lines of magnetic force that spiral out from the sun, the plasma radiates into space like a breeze from a fan. This breeze is the solar wind.

The solar wind radiates away from the sun in all directions, traveling between 700,000 and 1.5 million miles (1.1 million to 2.4 million km) per hour. It leaves the sun at a temperature of 1,800,000°F (1,000,000°C) and cools as it travels through space to 360,000°F (200,000°C). It is so thin that it transmits virtually no heat to our planet. As it rushes past Earth, the planet's magnetic field deflects it so that it flows around our planet just as water flows around a moving ship. But the wind also distorts Earth's magnetic field into a long teardrop shape, much like the wake left in the water behind the ship.

One visible effect of the solar wind is Earth's auroras. As the ions that compose the solar wind rush past Earth, they are attracted by the north and south magnetic poles, just as iron filings will gather around the ends of a bar magnet. As the particles spiral into the upper atmosphere, they strike molecules of oxygen, nitrogen, and other gases, causing them to glow, in the same way that electricity passing through the gas in a neon tube causes it to glow. We see these glowing gases as swirling rays and curtains of colored light, or auroras.

Earth's magnetic field acts as a kind of shield, protecting the surface of our planet from the radiation of the solar wind. Because of the ions that comprise it, the solar wind is electrical in nature. It would disrupt power grids and damage electronics if it could reach the surface of Earth. Its radiation would be harmful to humans and other animals. Particular powerful bursts of the solar wind, or coronal mass ejections, caused by storms on the surface of the sun, could damage satellites and even affect the surface of our planet. In 1989 a solar storm caused a blackout of the electric grid in the northeastern United States and Québec. The largest such storm ever recorded occurred in 1859. Brilliant auroras were seen all over the world, and telegraph lines not only failed but gave off sparks and shocked operators as the extra electrical energy poured into the wires.

MAGNETIC FIELDS

The sun and most of the planets, including Earth, act like giant magnets. If you were to take a rubber ball or a ball of clay and stick a bar magnet through its middle, you would have a model of Earth or the sun. The magnet is surrounded by a magnetic field, a region where energy as electrons flows from one pole of the magnet to the other.

This flow is represented by magnetic field lines. They describe the direction and shape of the flowing electrons. The field lines always follow a closed loop, connecting the opposite poles of the magnet.

You can see the field lines that surround a magnet by performing a simple experiment. Place a bar magnet under a sheet of paper. Sprinkle iron filings on the top of the paper. The bits of iron will outline the magnet's field.

Many things can distort the sun's magnetic field. The rotation of the sun causes it to spiral out like water from a lawn sprinkler, and the movement of the sun through space causes it to stretch out like the tail of a comet. The distortion of the magnetic field creates localized areas of strong magnetism, like knots in a tangled ball of string. These help create sunspots, solar flares, and coronal mass ejections.

THE BIG BUBBLE

The solar wind, part of the sun's atmosphere, forms a kind of huge bubble, or heliosphere, around the solar system. Although the word *heliosphere* suggests a symmetrical round shape, like a balloon, scientists don't know what shape it is. But most agree that the outermost limit of the heliosphere is at about 100 AU, more than twice the distance of the Kuiper belt.

Energetic events and objects in our galaxy, such as supernovas, black holes, and neutron stars (very dense, quickly rotating balls of neutrons that form when massive stars die) send radiation as particles racing through space at the speed of light. This radiation is galactic cosmic radiation, or cosmic rays. It contains a great deal of energy and could easily disrupt and damage delicate molecules such as DNA or even entire living cells.

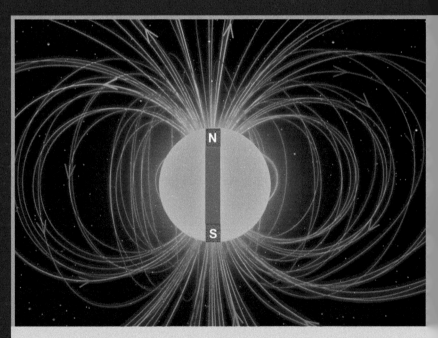

Magnetic field lines are invisible except for their effects on magnetic and electrically charged objects and particles. The direction of the magnetic field runs from the north pole to the south pole of the magnet.

The atmosphere and magnetic field of our planet helps shield life here from cosmic rays. But spacecraft and astronauts traveling beyond the atmosphere don't have this protection. They would be exposed to powerful cosmic rays that could easily harm delicate electronics or living cells. Fortunately, the heliosphere surrounding our solar system absorbs about two-thirds of these dangerous rays. Astronauts can be protected from the remainder by their spacecraft and space suit.

Just as a planet was like a ship with the solar wind flowing around it, the heliosphere is also like a ship. As the sun moves through space—carrying the solar system along with it—it moves through the interstellar medium, which fills the space between stars and galaxies. The interstellar medium is very thin, consisting of stray atoms of hydrogen and helium and a small amount of dust. The air around you has about 4×10^{20} atoms (about 400,000,000,000,000,000,000) in every cubic inch (16.4 cu. cm). But in interstellar space, only about sixteen atoms are in every cubic inch. That is four hundred thousand billion billion times thinner than the air you are breathing. To something as small as a spaceship, the interstellar medium may as well be a vacuum, devoid of any material at all, for all the effect it has. But to something as large as a planet—or as vastly huge as the heliosphere—its presence can be observed.

As the interstellar medium flows past the heliosphere, the effect is similar to the solar wind flowing past a planet. At about 100 AU, the solar wind has become so thin and weak that it can no longer resist the interstellar medium. The boundary between the heliosphere and the interstellar medium where the pressure between the two is balanced, is the heliopause, and because of its resemblance to a ship's bow wave, scientists also call it the bow wave. The point just past the bow wave, where the interstellar medium begins to stop the flow of the solar wind, is the termination shock. Just beyond the termination shock is the outermost region of the heliosphere, where the solar wind is slowed down by its collision with the interstellar medium. Ions pile up like traffic behind a slow-moving car, and it becomes denser and hotter. This region is the heliosheath.

Some scientists believe there may be a bow shock around the bow wave of the heliosphere. A bow shock is much like the shock waves that an aircraft creates as it exceeds the speed of sound in air. If the heliosphere moves fast

The Hubble Space Telescope captured this image of the bow shock around a neighboring star as it passes through the Orion Nebula, a region more densely packed with dust than the region around our sun.

enough through the interstellar medium, it will create a bow shock. But astronomers are still debating as to whether the bow shock exists.

To discover all of these things, astronomers need more than traditional telescopes. One reason is that the heliosphere can't be seen in visible light. It can only be measured with special instruments. One of these is the Daniel K. Inouye Solar Telescope in Hawai'i, a telescope that can detect and measure the sun's magnetic field. The telescope was specially designed to discover how the field is created and how it affects the space around it.

Some of the information astronomers have about the heliosphere comes from spacecraft that have traveled through the space beyond Pluto: the two Voyager spacecraft and New Horizons. Other spacecraft were sent in the opposite direction, to explore the source of the heliosphere: the sun. The Parker Solar Probe, launched in 2018, was designed to discover the origins of solar magnetic activity and the heliosphere. Many other probes have studied the sun, such as the Solar Dynamics Observatory and IBEX, which both float over Earth, and the STEREO mission, consisting of two spacecraft orbiting the sun. All of these have provided valuable information on how the sun's magnetic field is generated and the effects it has on the space around it.

CHAPTER 6

The Long Reach of the Sun

You can never get so far away from the sun that you escape its gravity. Isaac Newton's law of gravitation tells us that the effect of the sun's gravity falls inversely as the square of the distance from it. So, you divide the force of gravity by the square of the distance between the sun and the other object. Earth, for instance, is 1 AU from the sun. If it were to be moved twice as far away, to 2 AU, the force of the sun's gravity would be reduced by the square of the distance—it'd be four times less powerful. Pluto orbits at about 40 AU, so the sun exerts a pull on Pluto that is sixteen hundred times less than it does on Earth. And the outer limit of the Oort cloud is somewhere between 50,000 and 100,000 AU, so the force of the sun's gravity at that distance would be ten trillion times weaker than at the distance of Earth. Objects orbiting the sun at this distance are nearly halfway to the nearest star, Proxima Centauri, so there is a kind of tug-of-war between the two stars, with the outer limits of the Oort cloud between them.

As weak as it is at that distance, the sun's gravity can still be felt, and sometimes, like a fisher casting a very wide net, it can catch some very strange objects.

In October 2017 astronomers at the Panoramic Survey Telescope and Rapid Response System (Pan-STARRS) in Hawai'i, who make a special study of near-Earth objects, spotted what they at first thought was a faint comet. But as it approached the sun, it showed none of the characteristics of a comet. There was no tail or signs of gases and dust escaping as the sun warmed it. It eventually shot past the sun at a distance of 0.26 AU and a speed of 196,000 miles (315,431 km) per hour. When scientists calculated its orbit, they found that it wasn't elliptical, like those of comets and asteroids. Bodies following elliptical orbits will swing around the sun repeatedly. But the orbit of this object showed that it came plunging into the solar system from interstellar space and, after passing by the sun, headed back out into space, never to return.

The IAU designated it 1I/2017 U1. The *I* stood for "interstellar," the first time this distinction had ever been used. But the discoverers at Pan-STARRS named the object 'Oumuamua, a Hawaiian word meaning "scout" or "visitor from afar arriving first," and this name grabbed the public's attention. Although the mysterious visitor had a name, and although astronomers knew where it came from, no one knew exactly *what* the object was. For the first time, astronomers observed an interstellar object passing through the solar system, and it proved to be unusual.

Although it wasn't possible to get a detailed photograph of 'Oumuamua so astronomers could know what it actually looked like, they could tell from their data that it was reddish, like many asteroids and Kuiper belt objects. They also estimated its size and shape by measuring changes in its brightness as it rotated end over end every eight hours. Estimates for its length ranged from 1,300 to 2,600 feet (400 to 800 m). To explain the extreme variations in brightness, 'Oumuamua would have to be nearly ten times longer than it was wide. To their surprise, scientists realized that the object was shaped something like a hot dog or sausage.

Astronomers couldn't detect any heat radiating from the object, even at its closest approach to the sun. Scientists realized that it must be a very reflective object, reflecting more heat than it was absorbing, for it to remain so cold at such a close distance to the sun—at least ten times more reflective than a typical solar system asteroid.

Perhaps the strangest thing about the mysterious object was what happened as it was leaving the solar system: it *accelerated*. It picked up speed.

Astronomers were puzzled. If 'Oumuamua wasn't a comet and wasn't an asteroid, what was it?

AN ALIEN VISITOR?

'Oumuamua seemed to be so strange in every way that it was not long before someone said, What if 'Oumuamua is not a natural body like an asteroid or comet? What if it is artificial?

NASA launched Voyager 1 and Voyager 2 in 1977. In 2022 Voyager 1 was about 155 AU from the sun and Voyager 2 was about 129 AU. In forty thousand years, Voyager 1 will pass within 1.6 light-years of AC+79 3888, a star in the constellation of Camelopardalis and Voyager 2 will pass within 1.7 light-years of the star Ross 248. What if 'Oumuamua is similar to the Voyager spacecraft? What if it is a space probe from some distant alien world, drifting through the universe? If it were, it would explain many of its mysterious qualities, such as its strange shape and its sudden acceleration when leaving the solar system.

Some astronomers pointed out that the observed variations in 'Oumuamua's brightness could be explained by shapes other than the elongated sausage that was first proposed. 'Oumuamua might instead be disk-shaped, something like a fat pancake. Some even said, What if it were a very large, very thin disk? It could then act like a solar sail. Just as a sailboat is propelled by the wind, a solar sail would be propelled by a star's solar wind, or even by the pressure of the light itself. In 2019 the Planetary Society, a nonprofit that promotes space exploration, launched an experimental solar sail called LightSail 2 into orbit around Earth. The large, square sheet of silvered mylar plastic was not too different from the kind of plastic you might wrap food in. Even though LightSail 2 was only 344 square feet (32 sq. m), it was still large enough that the pressure of

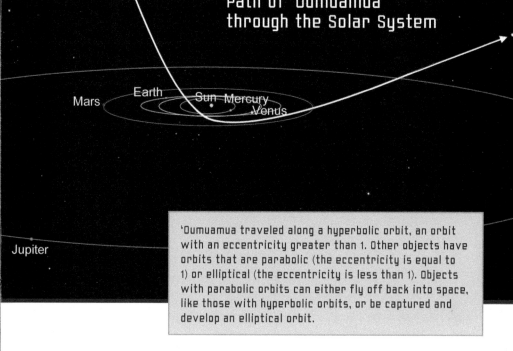

Mars

Earth

Sun Mercury

Venus

Jupiter

'Oumuamua traveled along a hyperbolic orbit, an orbit
with an eccentricity greater than 1. Other objects have
orbits that are parabolic (the eccentricity is equal to
1) or elliptical (the eccentricity is less than 1). Objects
with parabolic orbits can either fly off back into space,
like those with hyperbolic orbits, or be captured and
develop an elliptical orbit.

sunlight alone changed its orbit. A solar sail uses no fuel and will accelerate as long as sunlight or the solar wind reaches it. By constantly accelerating, a solar sail could eventually reach very high speeds. Many people believe solar sails are ideal for interstellar travel.

Perhaps, some scientists suggested, that's exactly what it was. 'Oumuamua wasn't an asteroid or comet. Instead, it was an interstellar probe, propelled by a giant solar sail, launched into space by a civilization on a planet of some distant star.

Other scientists were quick to point out flaws in this idea. Solar sails lose efficiency the farther they get from their main light source. They just won't keep going faster and faster forever. And 'Oumuamua was just too massive to be able to achieve any great speeds. It would be even slower than a conventional rocket. The top speed it could attain, they calculated, would be about 1,364 miles (2,195 km) per hour, about Mach 2—the speed of a fighter jet. Even at such a speed, it would still take two million years to travel the four light-years separating Earth from Proxima Centauri, the nearest star.

So probably 'Oumuamua was just a comet or asteroid—a very strange-looking comet or asteroid.

'Oumuamua may be an unusually shaped asteroid, as shown in the top image, or a solar sail developed by an alien civilization, as shown in the bottom image.

ANOTHER STRANGER

While 'Oumuamua was the first known interstellar visitor to the solar system, it wasn't the last. Two years after 'Oumuamua passed by, a Russian amateur astronomer, Gennadiy Borisov, spotted what he thought was a new comet. It was moving in an unusual direction, though, compared to how most comets that originate in the Kuiper belt or Oort cloud move. Astronomers in Poland and the Netherlands calculated its orbit and discovered that like 'Oumuamua, the new comet originated far outside our solar system, in interstellar space.

Astronomers discovered that the new comet, 21/Borisov, contained far more carbon monoxide than any comets known to orbit the sun. Such a composition meant the comet formed in a region of great cold. So, some scientists speculated that the comet had formed around a red dwarf star, a colder star than the sun. Red dwarf stars are the most common stars in our galaxy, and astronomers have observed higher concentrations of carbon monoxide around them as compared to other stars.

Unlike its predecessor, 21/Borisov showed a distinct coma and tail as it neared the sun, so there was no question that it was a comet. Its size was also typical for a small comet: a little over half a mile (1 km) wide.

The comet's swing around the sun wasn't as uneventful as 'Oumuamua's. As it approached the sun, astronomers observed a large piece of 21/Borisov break off. This happens to many comets. The frozen gases and ice comprising a comet's nucleus act as a kind of glue holding the comet together. As a comet nears the sun, more and more of the ice that holds it together turns to gas. When enough ice sublimates, not much is left to hold the comet together, and it starts to fall apart. Some comets have completely disintegrated as they passed the sun. As a comet's ice evaporates, much of it will escape the comet in geyserlike jets. These can act in the same way as the rockets that steer a spacecraft. They can start a comet spinning or even change the course of its orbit. Spinning like this might have contributed to 21/Borisov shedding bits and pieces of itself.

Jets of gas might also explain the sudden, mysterious acceleration of 'Oumuamua. Although scientists saw no evidence of gases being released from the object as it swung past the sun, unobserved jets may have acted like rocket boosters, increasing 'Oumuamua's speed.

This illustration shows what Comet 21/Borisov might have looked like as it broke apart.

CHAPTER 7

Between the Solar System and the Stars

The farther we get away from the sun, the more difficult it is to explore what is there. The objects beyond Neptune are small, dark, and far away. And some, like the solar wind, are invisible without special equipment. Yet it is important to discover what is there and how it got there. For instance, the heliosphere shields Earth from dangerous cosmic rays, radiation that would be harmful to life on our planet. How does it do this? Will it always protect us, or does it change over time?

The nearest stars are so far away that their light takes more than four years to reach Earth. Yet these stars have an effect on our solar system. They can, for instance, send comets hurtling into the inner solar system like missiles.

It was once easy to imagine our solar system—the sun, Earth, and the other planets—as a kind of island, separate from the rest of the galaxy, unaffected by whatever might be happening on distant stars. But astronomers have found that to be untrue. The solar system is part of a community. The gravity of distant stars tugs at the borders of the sun's domain. Black holes, supernovas, and neutron stars rain a constant torrent of radiation over our world. Even galaxies beyond our own Milky Way can affect our planet and the life that exists on it.

These discoveries have revealed just how important it is to know what goes on around our planet and solar system. But the space beyond the solar system is very difficult to study. Not only is it billions of miles away, but it is filled with things that are invisible and often hard to detect with even the most sensitive instruments.

EXPLORING THE DEEPEST REACHES OF SPACE

How *do* scientists learn about things so far away and so difficult to see and measure? Often the evidence is indirect. An otherwise invisible object may reveal its presence by the way it acts upon other objects. This was how Neptune and Pluto were discovered. Sometimes, scientists will infer the presence of an object or objects to explain a phenomenon. For instance, the presence of the Oort cloud was deduced because it explained where long-period comets came from.

Most of the objects that lay beyond Pluto are too small and dark to study with Earth-based telescopes. One alternative is to go there.

The two Voyager probes were the first to explore the space beyond Pluto. Launched back-to-back in August and September 1977, their primary mission was to explore Jupiter, Saturn, Uranus, and Neptune. This mission was completed in 1989 when Voyager 2 flew past Neptune. Since the spacecraft were still functioning perfectly, the NASA scientists overseeing them decided to give them a new mission: the Voyager Interstellar Mission. The mission's goals include studying the farthest limits of the sun's magnetic field and the solar wind.

The Voyager Interstellar Mission had three phases. The first measured the termination shock, where colliding with the interstellar medium slows

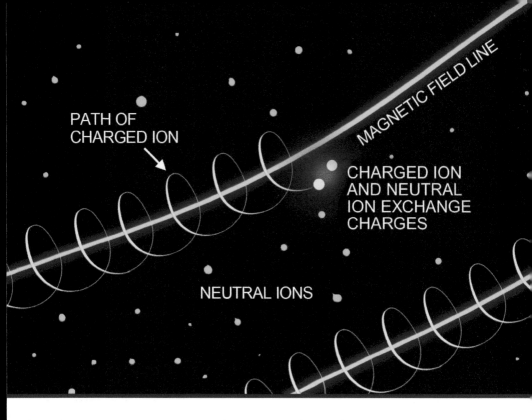

PATH OF
CHARGED ION

MAGNETIC FIELD LINE

CHARGED ION
AND NEUTRAL
ION EXCHANGE
CHARGES

NEUTRAL IONS

the solar wind from 1 million miles (1.6 million km) per hour to 250,000 miles (402,336 km) per hour.

Voyager 1 passed the termination shock in December 2004 and Voyager 2 in August 2007 at about 90 AU from the sun—twice the distance of Pluto. Then the spacecraft entered the heliosheath, the outer layer of the heliosphere. The spacecraft were nearing interstellar space, where the influence of the sun was barely felt any longer.

Finally, on August 25, 2012, Voyager 1 left the heliopause, becoming the first human-made object to enter interstellar space. It was 122 AU from the sun. Voyager 2 followed on November 5, 2018. Both spacecraft have enough power left to send information back to Earth until at least 2025. They will be nearly 130 AU from the sun by then, more than four times the distance of Pluto. As far as that is, it is still only one two-thousandth of the distance to the nearest star.

With the New Horizons spacecraft well into the Kuiper belt, it too has been returning valuable data about the outer solar system. This includes information about the nature of the solar wind at distances far from the sun.

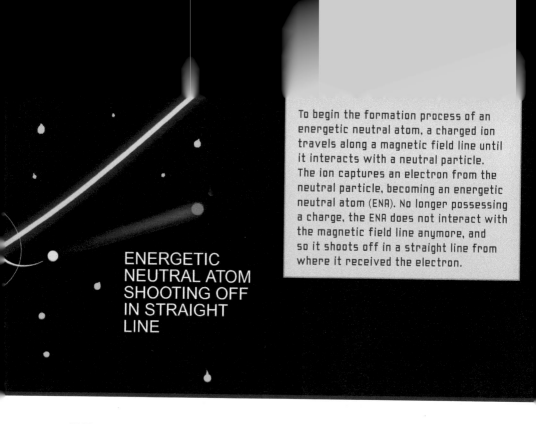

To begin the formation process of an energetic neutral atom, a charged ion travels along a magnetic field line until it interacts with a neutral particle. The ion captures an electron from the neutral particle, becoming an energetic neutral atom (ENA). No longer possessing a charge, the ENA does not interact with the magnetic field line anymore, and so it shoots off in a straight line from where it received the electron.

ENERGETIC NEUTRAL ATOM SHOOTING OFF IN STRAIGHT LINE

IBEX

Data from Voyager 1 and Voyager 2 suggested that the heliosphere was shaped like a bubble, but the spacecraft measured only two points on its surface. That's not nearly enough data to determine the shape of a three-dimensional space. In October 2008, NASA launched the Interstellar Boundary Explorer (IBEX) with the goal of more accurately mapping the boundary of the solar system.

The future of space travel is one reason for finding out as much as possible about the heliosphere. When spacecraft and humans travel outside the protection of Earth's atmosphere and magnetic field, they are exposed to cosmic rays. Although scientists know that the heliosphere shields the solar system from most cosmic rays, they still have some unanswered questions: Is the shielding provided by the heliosphere uniform? Is it the same everywhere? How much extra protection should we give to humans and electronics that travel into space? To answer these questions, scientists needed to find out what the heliosphere was shaped like. This was the goal of the IBEX mission.

To map the outline of the heliosphere, IBEX was equipped with a set of special telescopes. Instead of light, these telescopes look for energetic neutral atoms (ENAs). Neutral atoms—atoms with no electrical charge—from the interstellar medium enter the heliosphere and mix with charged particles in the solar wind. As a charged particle interacts with a neutral atom, the charged particle captures an electron and becomes neutral itself: an ENA. The speed of the particle doesn't change, but the particle is no longer affected by magnetic forces, and it travels in a straight line. Some of these travel toward the heliopause and bounce back toward the inner solar system, in the same way radar signals bounce off a distant object. By collecting and measuring these particles, IBEX was able to map the boundary of the heliosphere.

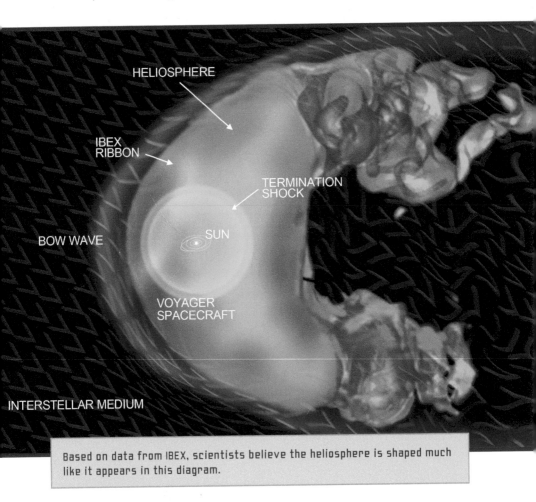

Based on data from IBEX, scientists believe the heliosphere is shaped much like it appears in this diagram.

MORE DISCOVERIES AND EVEN MORE MYSTERIES

Based on information provided by the IBEX satellite, the real shape of the sun's heliosphere was unexpected. Instead of the symmetrical bubble shape astronomers had expected and that the word *sphere* implies, the heliosphere seems to resemble a partly deflated, crescent-shaped balloon. There was no sign of the expected tail.

The shape and size of the heliosphere is not constant. It changes depending on the strength of the solar wind, which varies because of the solar cycle. Every eleven years, the sun's magnetic field flips and its north and south magnetic poles swap places. During the middle of a cycle, the sun reaches its maximum activity. Sunspots, solar flares, and coronal mass ejections increase. These can increase the number of auroras and affect electrical grids and communications on Earth. While the sun's activity is peaking, the solar wind becomes stronger. Its strength decreases again as the sun's activity wanes. This fluctuation affects the size of the heliosphere, just as blowing more air into a balloon makes it grow larger and letting the air out makes it become smaller.

But what about the strange crescent shape? Among the things measured by New Horizons was the presence of pickup ions in the heliosphere. When the neutral atoms from the interstellar medium have their electrons captured by charged particles, and the charged particles become ENAs, the formerly neutral atoms become ions themselves. These ions are then "picked up" by the magnetic field of the heliosphere and move along with it. Pickup ions are much hotter and more energetic than other solar wind atoms. The forces generated by the pickup ions prevail over the weaker solar wind atoms. All other things being equal, the outward pressure caused by pickup ions would create a spherical heliosphere, just as the pressure inside a balloon makes it round. But they also tend to quickly leak out of the termination shock, making the sphere deflate like a balloon losing its air.

Another factor affecting the shape of the heliosphere is the rotation of the sun. Since the solar wind is emitted by the sun and the sun rotates, the wind takes on a spiral shape as it moves away from the sun. So instead of inflating evenly like a balloon, the rotating solar wind and magnetic fields twist the heliosphere as it expands. This contributes to its odd shape.

The twisting of the magnetic field has another strange effect. As the sun spins, its magnetic field lines twist and cross over one another, like a tangled ball of yarn. And like a tangled ball of yarn, "knots" form. In the magnetic field, these become bubbles about 100 million miles (161 million km) across. Astronomers are especially interested in the effect such bubbles might have on future space exploration. They may make it easier for cosmic rays to penetrate the heliosphere—or, instead, they may trap cosmic rays, making it more difficult for them to reach the inner solar system. Researchers are hoping that future space probes may find the answer.

In addition to the bow wave, the termination shock, and the heliosheath, scientists had expected a bow shock where the heliosphere met the interstellar medium. The bow shock would have been equivalent to the sonic boom created when an aircraft exceeds the speed of sound. It occurs when the waves that form in front of the airplane can't move out of the way fast enough. Instead, they build up until they merge into a single, powerful wave, causing a loud boom. IBEX found that while a bow wave precedes the heliopause, there is no bow shock. It determined that the speed of the heliosphere through the interstellar medium is only 52,000 miles (83,686 km) per hour. If this measurement is correct, the heliosphere is too slow to create a bow shock, just as an airplane flying slower than the speed of sound won't create a sonic boom.

The shape of the heliosphere was not the last strange thing IBEX found. In searching for ENA emissions in 2009, the spacecraft detected a ribbonlike band of particles—later dubbed the IBEX ribbon—where emissions of ENAs were two or three times greater than in the rest of the sky. Scientists are not sure why the ribbon is there or what causes it. A popular hypothesis is that the ribbon exists in a special region of the heliosphere where ionized hydrogen atoms in the solar wind cross the magnetic field of the Milky Way galaxy. Since ions carry an electric charge, they are affected by the magnetic field. The atoms start to spin around the magnetic field lines. Some of the ions shoot off back toward the sun. This extra boost might be the cause of the ribbon.

While this theory seems to explain the ribbon, scientists still want to learn more about it and what causes it, as well as answer their further questions about the magnetic bubbles, what the strange shape of the

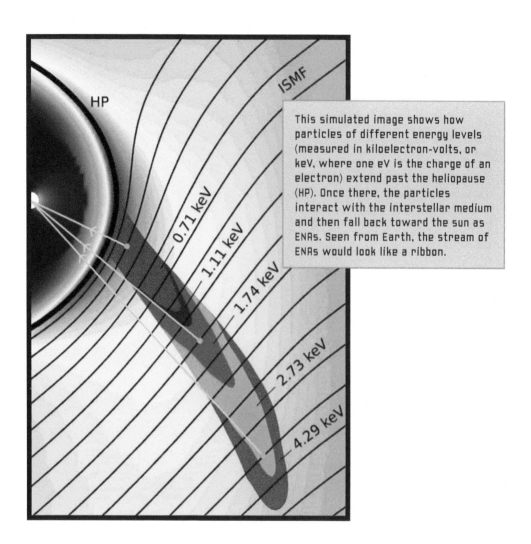

This simulated image shows how particles of different energy levels (measured in kiloelectron-volts, or keV, where one eV is the charge of an electron) extend past the heliopause (HP). Once there, the particles interact with the interstellar medium and then fall back toward the sun as ENAs. Seen from Earth, the stream of ENAs would look like a ribbon.

heliosphere means, how that shape came about, and what the future of the heliosphere might be. To help answer these questions, NASA has planned the Interstellar Mapping and Acceleration Probe (IMAP), scheduled to launch in 2024. IMAP will accurately map the boundaries of the heliosphere and help scientists better understand how it evolved, how it works, and how it interacts with the interstellar medium. The results of IMAP's mission are crucial, as the health of our planet and the future of space travel depend upon the protection the heliosphere offers.

CONCLUSION
DISTANT RELATIVES

The solar system has grown larger and larger the more science has learned about it. Thousands of years ago, humans believed that Earth was the center of the universe, that the world they lived on was all there was. Then they learned that it was one of several planets orbiting the sun. The number of planets grew and so did the size of the solar system. The solar system extends almost halfway to the nearest star. And it is composed of not only planets, asteroids, and comets but also powerful forces and vast fields of energy.

The outermost regions of the solar system are tens of billions of miles away from Earth, so far away that light itself may take two years to reach Earth from the farthest reaches of the Oort cloud. What could we learn from such distant, cold, dark places that could possibly relate to our own warm, comfortable planet?

The boundaries of the solar system are the remnants of the formation of the sun and its planets, frozen in time. After a house or building has been constructed, there will be a pile of leftovers: bricks, wood, wires, pipes, and other bits and pieces of the raw materials that went into the finished structure. That is what the Kuiper belt and Oort cloud are: leftovers from the beginning of our solar system. The ice and rock and dust orbiting there were once part of the primordial cloud in which our star formed. Studying what is found there is like traveling back in time more than four billion years to the birth of the sun and its planets—including Earth.

The solar system is like a single, large family. And to really understand and appreciate this family, it is necessary to discover where the family came from, what its origins are. When scientists look into the distant past to find out how our solar system came to be, it is very much like those scientists who study the origins of our species, *Homo sapiens*. The better we understand where we came from, the better we understand ourselves. Human beings are part of the planet Earth, which in turn is part of a vast system of thousands of worlds, large and small, extending halfway to the nearest star. Understanding how all that came to be helps us understand the origins of our planet, which then helps us understand ourselves.

To do this, scientists from around the world have joined forces, including observatories in North and South America, Europe, Asia, and Africa. Spacecraft have been sent beyond Earth to explore the sun's most distant reaches. Every year new, more powerful instruments are used by these explorers and new, more sophisticated spacecraft are proposed. China, for instance, is planning to launch a probe in 2024 that will travel deep into the heliosphere, 100 AU from the sun. The European Space Agency has plans for a probe that will reach 200 AU after a journey of 25 years and study the heliopause, while NASA hopes to reach the same distance in 2044 with its Innovative Interstellar Explorer. One of the most ambitious is NASA's Interstellar Probe. If approved, it would be launched in 2036 at a speed of 37,000 miles (59,546 km) per hour to reach the termination shock in merely 12 years. It would then keep going, perhaps reaching 1,000 AU. As incredible as that distance is, it is still only three-hundredths of the distance to the nearest star, and the Interstellar Probe would take 150 years to get there.

With these new instruments, with each new space probe, scientists learn a little more about the nature of the sun, the solar system, and the world we live in.

With every new mystery that is solved, with every question that is answered, a hundred more arise. But that is how science works and what makes it so interesting and so exciting. There is always something brand new to learn.

GLOSSARY

accretion: when an object grows by slowly accumulating matter

astronomical unit (AU): the average distance of Earth from the sun, approximately 93 million miles (150 million km)

aurora: the glow of gases in the upper atmosphere of a planet caused by the impact of the solar wind

barycenter: the center of gravity around which two bodies orbit

big bang: the event that marks the beginning of the observable universe, in which the universe began as a single point, then expanded rapidly

binary planet: two bodies orbiting a common center of gravity

bow shock: the hypothetical shock wave where the heliosphere meets the interstellar medium

comet: an icy interplanetary body that releases gases that form a bright head and tail when it approaches the sun

convection: movement in a gas or liquid where hot or warm material rises and colder material sinks

corona: the outermost layer of the sun's atmosphere, visible from Earth only during a solar eclipse or with special instruments

coronal mass ejection: an explosive release of plasma and electromagnetic energy from the sun

cosmic ray: a high-energy atomic particle that originates in supernovas and other energetic sources

cryolava: very cold liquid flowing from a cryovolcano

cryovolcano: a volcano that erupts cold liquids and ice instead of molten rock

dust: very small particles of matter ranging in size from a few molecules to 0.1 mm (0.004 inches)

ecliptic plane: the imaginary disk that stretches from the sun's equator through the solar system. Most of the planets and the Kuiper belt lie on this plane.

electron: an atomic particle carrying a negative electric charge

energetic neutral atom (ENA): a fast-moving particle with a neutral electric charge

fusion: when a new atom is created by the combination of two simpler ones, resulting in the release of large amounts of energy

galactic cosmic radiation: high-speed atomic nuclei emitted by supernovas, black holes, and neutron stars

gas giant: a planet dominated by gases such as hydrogen and helium

greenhouse effect: the heating of a planet surface when infrared radiation is trapped beneath the atmosphere

heliopause: the boundary between the solar wind and the interstellar medium where the pressure between the two is equal

heliosheath: the area between the termination shock and the heliopause

helium: the second-lightest element and the second most abundant in the universe (after hydrogen)

hydrocarbon: a molecule made entirely of carbon and hydrogen atoms

hydrogen: the most abundant, simplest, and lightest element in the universe. It consists of a single proton and electron.

hydrostatic equilibrium: when the gravitational pull of a planet or star exactly matches its internal outward pressure

infrared: a wavelength of light beyond red in the visible spectrum. The heat you feel when you are out in the sun or standing in front of a fire is infrared radiation.

interstellar medium: the thin gas (mostly helium) and dust that fills the space between stars

ion: an electrically charged atom or molecule

ionization: the creation of ions when an electron is gained or lost by an atom

Kuiper belt: a disk-shaped region of space beyond Neptune with many large icy bodies

Kuiper belt object: any of the many small, mostly icy bodies orbiting the sun outside the orbit of Neptune

Late Heavy Bombardment: a period about four billion years ago when a large number of asteroids collided with Earth and its moon

long-period comet: a comet that takes longer than two hundred years to orbit the sun

magnetic field: the region of energy surrounding a magnet

magnetic field lines: a representation of the flow of energy from one pole of a magnet to the other

methane: an odorless, colorless gas made entirely of carbon and hydrogen atoms. Its chemical formula is CH_4.

molecule: a combination of two or more atoms

Oort cloud: a spherical region of space containing many icy objects that lies between 2,000 and 100,000 AU from the sun

organic molecule: a complex molecule that contains carbon atoms

perihelion: the point in the orbit of a planet or comet when it is closest to the sun

planet: a solid or partially liquid body orbiting a star, too small to trigger nuclear reactions in its core but large enough to be spherical due to hydrostatic equilibrium

plasma: atoms that have been heated to a temperature that has stripped them of one or more of their electrons. It is a state of matter along with solid, liquid, and gas.

polygon: a geometrical shape with many sides, such as a square or hexagon

primordial black hole: a hypothetical type of black hole created in the first second after the big bang

protoplanetary disk: a disk of dust and gas surrounding a star that will eventually form into planets

protostar: a cloud of gas and dust dense enough for gravitational collapse to begin

pseudoscience: something that appears to be science but does not follow scientific methods

short-period comet: comets that take less than two hundred years to orbit the sun

solar flare: an eruption from the sun similar to a coronal mass ejection but smaller

solar system: the sun and all the planets, comets, asteroids, and other bodies that orbit it

solar wind: the stream of gas flowing rapidly from the sun past Earth and other planets

sublimation: when ice turns into a gas without first going through a liquid phase

supernova: a star that explodes, blowing off most of its mass, leaving only its dense core

termination shock: the region where the solar wind slows to less than the speed of sound

terrestrial planet: a planet, like Earth, composed mostly of iron and rock

thermal erosion: erosion caused by changes in temperature

SOURCE NOTES

19 Tony Simon, *The Search for Planet X* (New York: Scholastic Book Services, 1962), 47.

41 Thomas H. Maugh II and Alan Zarembo, "Fred L. Whipple, 97; Scientist Found Comets Are 'Dirty Snowballs,'" *LA Times*, September 1, 2004, https://www.latimes.com/archives/la-xpm-2004-sep-01-me -whipple1-story.html.

54 Ron Miller, *Mercury and Pluto* (Minneapolis: Twenty-First Century Books, 2003), 72.

SELECTED BIBLIOGRAPHY

Bradford, Cannon. "A Discussion on Interstellar Pickup Ions." University of New Hampshire. Accessed January 1, 2022. https://ceps.unh.edu/sites/default /files/2013_seniorthesis_cannon.pdf.

Krajnović, Davor. "The Contrivance of Neptune." arXiv, October 28, 2016. https://arxiv.org/ftp/arxiv/papers/1610/1610.06424.pdf.

"Naming of Astronomical Objects: Dwarf Planets." International Astronomical Union. Accessed January 1, 2022. https://www.iau.org/public /themes/naming/#dwarfplanets.

"The Rosetta Lander." European Space Agency. Accessed January 1, 2022. https://www.esa.int/Science_Exploration/Space_Science/Rosetta/The _Rosetta_lander.

Simon, Tony. *The Search for Planet X*. New York: Scholastic Book Services, 1962.

Talbert, Tricia. "Five Years after New Horizons' Historic Flyby, Here Are 10 Cool Things We Learned about Pluto." NASA, July 14, 2020. https:// www.nasa.gov/feature/five-years-after-new-horizons-historic-flyby-here-are -10-cool-things-we-learned-about-plut-0.

———. "The Icy Mountains of Pluto." NASA, July 15, 2015. https:// www.nasa.gov/image-feature/the-icy-mountains-of-pluto.

Wall, Mike. "Is Our Solar System Shaped like a Deflated Croissant?" Space.com, August 7, 2020. https://www.space.com/solar-system-heliosphere -shape-croissant.html.

"What Was the Carrington Event?" SciJinks. Accessed January 1, 2022. https://scijinks.gov/what-was-the-carrington-event/.

FURTHER INFORMATION

Books

Brown, Mike. *How I Killed Pluto and Why It Had It Coming*. New York: Spiegel & Grau, 2010.
Planetary astronomer Mike Brown helped discover many Kuiper belt objects, including Eris, which led him to instigate the process of reclassifying Pluto as a dwarf planet. This book covers his account of events surrounding the debate and why he made this decision.

Chambers, John, and Jacqueline Mitton. *From Dust to Life: The Origin and Evolution of Our Solar System*. Princeton, NJ: Princeton University Press, 2017.
This book is a must-read for learning more about the formation of the solar system. The authors draw on the latest research and discuss new controversies and debates. It also covers how Earth became a prime candidate for the evolution of life.

Dethloff, Henry C., and Ronald Schorn. *Voyager's Grand Tour: To the Outer Planets and Beyond*. Old Saybrook, CT: Konecky & Konecky, 2009.
A sweeping account of the Voyager missions, this book covers the probes' construction, launching, and findings.

Matloff, Gregory L. *Deep Space Probes*. New York: Springer, 2005.
Check out this fascinating book to learn more about all the probes sent from Earth to the farthest regions of the solar system.

McGinty, Alice P. *The Girl Who Named Pluto: The Story of Venetia Burney*. New York: Schwartz & Wade, 2019.
This picture book provides one of the most thorough accounts of Venetia Burney, who came up with the name Pluto for the newly discovered ninth planet.

Miller, Ron. *Natural Satellites: The Book of Moons*. Minneapolis: Twenty-First Century Books, 2021.
The solar system is full of unusual and interesting objects, and many of them orbit the major planets. Miller walks through the moons of each planet, explaining their unique properties and which ones may even have life.

Siddiqi, Asif A. *Beyond Earth: A Chronicle of Deep Space Exploration, 1958–2016*. Washington, DC: NASA History Program Division, 2018.

This book chronicles space exploration over the past century, beginning with missions to the moon and extending to Mars, Saturn, the Kuiper belt, and beyond. Focusing on robots, Siddiqi profiles probes, rovers, and satellites, explaining their technical aspects and mission goals.

Stern, Alan. *Chasing New Horizons: Inside the Epic First Mission to Pluto*. New York: Picador, 2018.
  Astronomer Alan Stern is the New Horizons project's principal scientist. Read his account to learn about the scientists who worked on the mission, the political and economic issues they faced, and the power of teamwork and scientific discovery.

———. *The Pluto System after New Horizons*. Tucson: University of Arizona Press, 2021.
  As a result of the New Horizons data, Stern wrote this book to provide the most thorough and up-to-date look at Pluto and its moons. While intended for fellow astronomers, this book is crucial for any Pluto enthusiast.

Toms, Ron. *Welcome Back, Pluto*. San Antonio: RLT, 2021.
  Author Ron Toms details the case in favor of Pluto's classification as a planet, not a dwarf planet per the IAU definition.

Tyson, Neil deGrasse. *The Pluto Files: The Rise and Fall of America's Favorite Planet*. New York: W. W. Norton, 2009.
  With his characteristically witty prose, celebrity astronomer Tyson explains why so many astronomers no longer consider Pluto a planet, why so many people became upset with the reclassification, and how the debate became such a prominent cultural event.

Magazines

Astronomy
  https://astronomy.com/
  An excellent magazine about astronomy for both the beginner and expert. Articles cover everything from the history of astronomy to the latest discoveries.

Sky & Telescope
  https://skyandtelescope.org
  Although aimed toward more experienced students and astronomers, its articles are easy for beginners to read too.

Websites

Astronomical League

https://www.astroleague.org/astronomy-clubs-usa-state
Here is a list of astronomy clubs in the United States. Clubs like these have special programs and lectures and host regular "star parties" where members gather on clear nights to share their telescopes and observe the moon, planets, and stars. You don't need to own a telescope to join a club. All you need is an interest in astronomy.

IBEX

https://www.nasa.gov/mission_pages/ibex/index.html
The official site for the IBEX mission provides a closer look at the spacecraft and the science of the heliosphere.

New Horizons News Center

http://pluto.jhuapl.edu/News-Center/index.php
Learn about the New Horizons mission that passed by Pluto and showed us that the planet is pink! The official site for the mission features the spacecraft's photographs and more information about the discoveries astronomers made from the mission.

Planetariums

https://www.go-astronomy.com/planetariums.htm
Here is a list of all the permanent planetariums in the United States and the world. Planetariums are an excellent way to learn about astronomy and also have fun.

Solar Dynamics Observatory

https://www.nasa.gov/mission_pages/sdo/main/index.html
The Solar Dynamics Observatory orbits Earth and monitors the sun for activity. Updates about the spacecraft and new findings about the sun are on the mission's main page.

Voyager Mission Status

https://voyager.jpl.nasa.gov/mission/status/
Keep track of where Voyager 1 and Voyager 2 are venturing next. This website provides their current status and news about the two probes.

INDEX

PHOTO ACKNOWLEDGMENTS

Images by Ron Miller. Additional images by NASA/ESA/Zena Levy, STScI, p. 4; ESA/Hubble and NASA, R. Cohen, p. 12; ALMA (ESO/NAOJ/ NRAO), S. Andrews et al., NRAO/AUI/NSF, S. Dagnello, pp. 14–15; nivium/Wikipeda (CC 2.0), p. 22; U.S. Naval Observatory/NASA, p. 24; NASA/JHUAPL/SwRI, p. 29; NASA/JHUAPL/SwRI, pp. 30–31; NASA/Johns Hopkins University Applied Physics Laboratory/Southwest Research Institute/Alex Parker, p. 32; Davide Pischettola/NurPhoto/Getty Images, p. 41; ESA/Rosetta/NAVCAM (CC BY-SA 3.0), p. 62; NASA/Carla Thomas, p. 65; NASA/ESA, p. 69; NASA/SwRI/Zirnstein, p. 85.

ACKNOWLEDGMENTS

The author appreciates the generous help and advice of

Dr. William K. Hartmann, senior scientist of the Planetary Science Institute

Dr. Alan Stern, principal investigator of the New Horizons mission

Dr. Philip Metzger, planetary physicist with the planetary science faculty at the University of Central Florida

Dr. Merav Opher, professor of astronomy at Boston University

ABOUT THE AUTHOR

Ron Miller is the award-winning author and illustrator of more than seventy books. In addition to the art he creates for his own books, Miller has illustrated dozens of others, including the best-selling *Zoomable Universe*. He is also a regular contributor to magazines such as *Astronomy* and *Scientific American*. He has been a production designer and illustrator on films such as *Dune* (1984) and *Comet Impact* (2000) and a participant in the NASA Fine Arts Program. Miller's Pluto postage stamp, attached to the New Horizons spacecraft, is now deep within the Kuiper belt, having traveled farther than any other stamp in history.